USS HORNET

USS HORNET

CV-12 SERVICE IN WAR AND PEACE

RUSSELL MOORE

America Through Time is an imprint of Fonthill Media LLC
www.through-time.com
office@through-time.com

Published by Arcadia Publishing by arrangement with Fonthill Media LLC
For all general information, please contact Arcadia Publishing:
Telephone: 843-853-2070
Fax: 843-853-0044
E-mail: sales@arcadiapublishing.com
For customer service and orders:
Toll-Free 1-888-313-2665

www.arcadiapublishing.com

First published 2023

ISBN 978-1-63499-442-2

Typeset in Sabon 10.5pt on 13pt
Printed and bound in England

Preface

This is a book about a large, mostly metal, object. This seems to be a funny way to describe the *Hornet*, but it is important to remember when reading this book. This object happens to have been formed into a large ship, designed to operate aircraft. Because of its design, construction, operation, and timing, this large ship turns out to be one of the most historic in the US Navy.

The *Hornet* is an incredible ship that has served the public of the United States for going on eighty years (as of the writing of this book). To tell this story I have tried to provide the appropriate level of context. What was the time that the ship served in? What was happening in the United States and abroad? What were the people on the ship doing? What directly happened to and on the ship? What happened around the ship? Once you combine the skill, determination, courage, and luck of the crew with the engineering of the ship into a story, it turns out to be an amazing one.

My interest in the war in the Pacific started when I was around eleven years old. It was there that I first obtained a copy of and read Robert D. Loomis' *Great American Fighter Pilots of World War II*. As much as I liked the whole book, my deepest curiosity was held by the Pacific War, especially the parts about the Cactus Air Force and the aircraft carriers. This led me to decades of increased interest in World War II and study on my own time.

I started working at the USS *Hornet* Museum in 2019. For me it was a perfect match. I was able to work somewhere where my interest in history, military history, and space exploration all came together in one job. I thought I knew a lot about World War II and aircraft carriers, but I realized as soon as I started that I had so much more to learn. This education began on my first day as I started taking docent led tours right away.

There were books to read, videos to watch, docents to talk to, and veterans to interview. I was able to meet and talk to admirals, astronauts, and aircraft pilots. There were people who served in World War II, the Korean War, the Vietnam War, the Cold War, and at NASA. Luckily for me, the *Hornet* museum also has a great library of donated books that I was able to use for my research.

For eighty years, the USS *Hornet* has served this nation, adapting every time she needed to for whatever mission was assigned to her. Designed before World War II even started, the *Hornet* reflected the best information about aircraft carriers at the time. During the war, she fought with distinction and the main modifications during the war were to her anti-aircraft guns and air group composition. Behind the scenes there were improvements to radar, communications and of course the handling of large groups of aircraft. By the end of World War II, the thinking on aircraft carriers had changed somewhat but the basic design proved incredibly useful. Only with the introduction of the *kamikaze* did the *Essex* carriers start to display some weaknesses. Even then, the lack of an armored flight deck was compensated for by the increasingly large anti-aircraft batteries and the addition of extra fighter aircraft.

After the war, the *Hornet* quickly transformed into a large troop transport and then back again. After some time out of commission (since the US Navy was simply too big), she came back into service modified to fly jet aircraft for the Korean War although she was too late to participate. She did serve throughout the Cold War.

Over the next few years, the *Hornet* was modified again and again, gaining an angled flight deck and eventually transforming into a specialized anti-submarine warfare aircraft carrier. This configuration served during the Vietnam War and turned out to be a perfect platform to participate in the Apollo program. The *Hornet* picked up the first and second sets of astronauts to walk on the moon, cementing her legacy as one of the all-time historic ships of the US Navy.

This was followed by years in the mothball fleet and in the end, a trip to the scrapyard. However, this trip was interrupted and instead the *Hornet* was saved and given one last mission as a ship museum. Now her role is changed to one of education and historical preservation. This is probably her last mission but it has been ongoing for twenty-four years and counting. Only time will tell how long this venerable ship will keep serving the public.

I am happy to have a chance to tell the story about this ship and hopefully I have added some new information and some new perspectives. I wanted the story to be one of adaptation and perseverance, two traits that the *Hornet* symbolizes.

Acknowledgments

I would like to first of all acknowledge the publishers and editors at Fonthill Media LLC. They gave me the opportunity to write my first book and their advice and suggestions have proved invaluable.

I would also like to acknowledge all the authors before me who wrote books about *Essex*-class carriers, the *Hornet*, World War II, the Cold War, the Vietnam War and the Apollo program. I read a lot of books on these topics and each of them helped to add to my understanding of the long and amazing history of the *Hornet* and the decades she served in.

I also want to acknowledge the *Hornet* crew members that I was able to interview personally (Dale Berven, Walter Davis, Bruce Douglas, Joe Holt, Joseph Leather, Chaplain Wallace Whatley, and Gene Millen) as well as the interviews that others have done that I was able to watch on video. There are dozens more *Hornet* crew members that contributed in small ways in conversations I would have with them on the ship. I am sure everyone must have gotten tired of all of my questions at some point.

Finally, I could not have completed this book without my wife Wahida's support. She let me take up valuable hours when we could have been having fun together to spend time researching and writing this book. She was always full of encouragement and had more faith that I could actually write a book than I did sometimes.

Contents

1

Design and Construction

Design

The design of the *Essex*-class aircraft carriers began in 1939. This was happening while the *Yorktown*-class USS *Hornet* (CV-8) and the *Wasp*-class USS *Wasp* (CV-7) were still under construction. Starting with the first US carrier, USS *Langley* (CV-1), which was converted from the collier USS *Jupiter* in 1920, the US Navy had gained a lot of experience in aircraft carrier operations. What they had not gained was combat experience. The design of the *Essex* class was going to have to be done with the best understanding that the US had without the benefit of wartime conditions. The *Essex*-class carriers were specifically built with the mission of fighting the Japanese in the Pacific. While under construction, some of the lessons that were being so harshly learned in the Battles of Coral Sea, Midway, and the fighting in the Solomon Islands were incorporated to the extent that they could. Mostly this resulted in changes in anti-aircraft armament.

The navy was expanding in 1939 as war was starting around the globe. The US knew it was going to be fighting soon, perhaps on two fronts against Germany and Japan. Congress had passed multiple Acts increasing the size of the navy in the 1930s: "The Vinson-Trammell Act of 1934" and "The Naval Act of 1938." The second act was a direct response to Japanese fighting in China and the German annexation of Austria and led directly to the beginning of the design process for the *Essex*-class carriers. This act allowed for the navy to begin building bigger ships (which had technically been allowed but not acted on since 1936 when the Italians and Japanese backing out of the revisions to the Washington Naval Treaty of 1922). The navy decided to build one more *Yorktown*-class carrier (USS *Hornet*) and then began the design of their dream aircraft carrier in the *Essex* class. The "Two-Ocean Navy Act" was passed on July 19, 1940 (signed into law after the fall of France) and this dramatically increased the number of *Essex*-class carriers that would be built—up to eighteen in the original act.

USS *Hornet* (CV-12) on the Elizabeth River, November 29, 1943.

The forces required for the fighting in the Pacific Ocean would be very different from that in Europe/the Atlantic. Although many have claimed that the losing of the US battleships at Pearl Harbor forced the US Navy to switch to a carrier-centric force, there were some who were already starting to consider the transition from battleships to aircraft carriers as the primary capital ship of the future. Representative Carl Vinson (D-GA) said in June 1940 during discussion of the Two-Ocean Navy Act that "the modern development of aircraft has demonstrated conclusively that the backbone of the Navy today is the aircraft carrier." It was still a time of change, however, and the number of battleship-first advocates was quite large.

The *Essex*-class carriers were essentially going to be updated *Yorktown*-class carriers. Two of these ships (USS *Yorktown* CV-5 and USS *Enterprise* CV-6) were active and had been running exercises in the 1930s testing out aircraft carrier principles along with their predecessors (the *Langley*-class USS *Langley* (CV-1), *Lexington* class including the USS *Lexington* (CV-2) and the USS *Saratoga* (CV-3) as well as the *Ranger*-class USS *Ranger* CV-4). This included several successful surprise attacks on Pearl Harbor and the Panama Canal in naval exercises. All of this information was brought to bear in designing the *Essex* class. Some of the major decision points were how big the ships could be, how many aircraft could they hold, how big was the flight deck going to be, what armament was needed, where would the elevators be, what and where to armor the ship and more. Complicating matters were that the ships had to fit in the Panama Canal and that naval aircraft were getting larger and heavier. Finally, the navy wanted the flight deck to be large enough to hold and launch all of its aircraft.

In the earlier aircraft carrier designs, the navy had learned a lot. The USS *Langley* (CV-1) was a converted collier launched in 1920. It was the smallest aircraft carrier the US built until the 1940s (the *Independence*-class light aircraft carriers and the various classes of escort aircraft carriers were smaller). It was the shortest, slowest (with only had three turbo-electric boilers) and held the least aircraft, but the navy had to start somewhere. During the early 1920s, the navy had many firsts on the *Langley*, testing out various ways to launch and land aircraft and many other aspects of air operations at sea.

Next to be constructed were the *Lexington*-class CVs USS *Lexington* (CV-2) and USS *Saratoga* (CV-3) launched in 1925. Due to the navy treaties at the time the US Navy had to stop construction of two battlecruisers and instead was able to convert them to aircraft carriers to stay within treaty requirements. These two carriers were the largest, most heavily armed and armored that the navy built before the end of World War II. They had cruiser armament (eight 8-inch guns and twelve 5-inch guns) and battlecruiser armor. They also had massive power: sixteen turbo-electric boilers (180,000 standard horsepower) allowed these ships to travel at 33 knots! With the ability to carry up to seventy-eight aircraft, for the first time these ships allowed the navy to practice large air squadron exercises. These two ships were over 300 feet longer than the *Langley* and carried six times as much crew. They were truly the US Navy's first fleet carriers.

The USS *Ranger* (CV-4), the first purpose-built aircraft carrier, launched in 1934. It was in between the *Langley* and *Lexington* classes as far as displacement, speed, and size. The *Ranger* was a step forward in many ways as, despite its smaller size, it had three elevators and three catapults (one more than the *Lexington* class of each) and could still handle almost the same number of aircraft. With only six boilers and two steam turbines, it was slower (29 knots) but not significantly. One of the reasons the *Ranger* was smaller was due to existing treaties that limited naval tonnage. This forced the designers to think differently about how an aircraft carrier should be designed. Gone were the heavy armor plate and 8-inch guns of the *Lexington* class. Instead only key areas were armored (and the armored belt was much thinner) and only eight 5-inch guns were used.

The next class to be designed and built was the *Yorktown* class. This class of ship was truly the starting template for the *Essex* class. The USS *Yorktown* (CV-5), launched in 1936, the USS *Enterprise* (CV-6), launched in 1938, and the USS *Hornet* (CV-8), launched in 1941, rounded out this class. All the ships of the *Yorktown* class were over 20,000 tons and at least 760 feet long. They carried up to ninety aircraft, were fast (32.5 knots), had a lot of power (nine boilers and four steam turbines generating 120,000 standard horsepower), and had similar armament and armor to the *Ranger* class. Still under the restrictions of the naval treaties of the day, these carriers never the less represented most of what the navy wanted at the time.

The final class designed and built before the *Essex* class was the USS *Wasp* (CV-7) of the *Wasp* class, launched in 1940. This was a smaller carrier like the *Ranger* and was built this way to utilize the remaining tonnage allowed under the soon to be discarded naval treaties. The *Wasp* was more like the *Ranger* in displacement, although it was unique in being the shortest aside from the *Langley*. It still held a large number of aircraft, three elevators (including an innovative, folding, side elevator), the same engines, boilers, armor, and weaponry as the *Ranger* class.

The designers of the *Essex*-class carriers had these eight aircraft carriers (two still under construction) of the US Navy, plus what they could learn about the aircraft carriers of the British and Japanese navies to go by. The only true war experience any navy had with their aircraft carriers was Japan's in their war against the Chinese. That did not help much as there was not any practical opposition to the Japanese Navy by the Chinese. Even when World War II started in Europe, there were no aircraft carrier clashes to study. The final design was approved in February 1940 and this was before some of the famous use of aircraft carriers by the British such as the attack on the Italian fleet at Taranto in November 1940 and the hunt for the *Bismarck* in May 1941.

Once the navy decided they needed to design a new type of aircraft carrier, they created a team led by Commander Leslie Kniskern. Commander Kniskern was an experienced naval designer who graduated from the US Naval Academy in 1922 and joined the Naval Construction Corps. He later did graduate level work (including at *Ecole d'Application du Genie Maritime* in Paris) and over his career was responsible for multiple ship designs. The navy brought in many experts to join his team but perhaps the most important of all was Commander James Russell, an aviator. Commander Russell graduated from the Naval Academy in 1926 and began aviator training in Pensacola Florida in 1928. He also completed graduate level work (in aeronautical engineering). By the time that the *Essex*-class carrier was being designed, Commander Russell had actually landed on all six active aircraft carriers. This made him uniquely valuable in representing the needs of the naval aviators.

To try to understand how complicated the ship would have to be, you have to imagine that an aircraft carrier has to be both a ship and an airport. It has to carry its own fuel and ammunition as well as fuel and ammunition for its aircraft. It also has to be a small city to feed and house thousands of sailors. This includes things like having a barbershop, tailor, cobbler, post office, mess halls, and much more. The ship has to be large and fast because it needs to have strong wind across the bow to help planes launch. This generally meant the ship had to be able to make around 30 knots. It has to be stable and structurally sound while having large open areas for the aircraft to be serviced and stored. Different power plant options were considered as well as different armored configurations-including having a heavily armored flight deck. Six major designs were considered, A–G.

The design that was ultimately accepted was a compromise of sorts. They used many features of the *Yorktown* class but included innovations from the other classes as well. Differences of opinion between the aviators and the naval

USS *Hornet* (CV-12) departing Norfolk Navy Yard after commissioning, November 29, 1943.

architects were ironed out. The design met all the specifications that the navy has asked for and was approved on February 21, 1940.

For the flight deck, the two major considerations were whether to make it armored or not and how big it needed to be. Related to that was the shape of the deck. The decision about whether or not to armor the flight deck was not an easy one. The supporting structure needed to hold an armored flight deck would make the hangar bay smaller which in turn would mean less aircraft could be carried. It would also mean that the design of the hangar bay would have to be less open, reducing ventilation (a serious concern on a ship carrying large amounts of flammable materials). The ship would also be top heavy—a problem for overall stability. On the other hand, it would make the ship much tougher to damage via dive-bombing (although if damaged the flight deck would be hard to repair without a shipyard or floating dry-dock). The unarmored design was not strictly unarmored, it simply had much thinner armor, 1.5 inches topped by a 3.5-inch teakwood deck. The advantages were several, however: a larger, more open hangar bay let the ship carry and service more planes; it allowed for better ventilation; and in combat, it was easier to push burning debris and aircraft over the side via the large side doors that could be opened. The teak flight deck could be repaired quickly even in combat (as evidenced by the USS *Yorktown* (CV-5) at Midway). This design was ultimately accepted and the main armor plate was put on the hangar bay (2.5 inches) deck to protect the vital areas below (especially the engine and fire rooms) and this helped to keep the ship more stable.

The flight deck had to be large to accommodate the idea of the "Sunday Punch" which was the US Navy's concept that an aircraft carrier had to be able to hold all its planes on deck and still have the capability to launch them. The flight deck was larger than the *Yorktown* class given the understanding that naval aircraft continued to grow. Bigger aircraft took up more space in the hangar bay and on the flight deck and needed more room to take off. In the end, an 846-foot-long flight deck was used (106 feet wide). A big design innovation was to keep the front of the flight deck square instead of tapering it off. The argument between the aviators and the designers was that keeping it squared off made it easier for planes to launch while tapering it off made it more structurally sound *v.* inclement weather. The aviators won this argument thanks to Commander Russell. In 1945, when hit by a typhoon, two of the *Essex*-class carriers, including *USS Hornet* (CV-12), had the front part of their flight decks partially collapse, somewhat vindicating the designer's fears.

An important part of the flight deck and the hangar bay are the elevators and the catapults. Based on the success of the USS *Wasp* (CV-7) with a foldable, side elevator, it was decided to have one of the three elevators on the *Essex* class built the same way. This enabled these ships to fit through the Panama Canal by folding the elevator against the side of the ship but also allowed for superior aircraft handling that comes with having three elevators. Elevator one was near the bow, allowing aircraft to be brought up from the front of the hangar bay to the launch area. Elevator two was the folding one on the port side across from the island which was accessed through a side door of the hangar bay. This elevator was well positioned to support bringing landed aircraft down and launching aircraft up. Because it was on the side, its position did not impact the space in the hangar bay. Elevator three was aft of the island and could quickly bring landed aircraft down to the hangar bay. However, its position when up left an area in the hangar bay that was unusable because of the large elevator well. The first two elevators were 48.2 feet by 44.2 feet and could hold 28,000 pounds. The third (side) elevator was the same size but was only able to carry 18,000 pounds.

Several catapult configurations were tried as the *Essex*-class carriers were being designed and built. The USS *Hornet* (CV-12) for example had one type H mark IV hydraulic catapult on the port side of the flight deck and one across the hangar bay to launch aircraft directly from the hangar deck. This catapult was very rarely used and eventually removed from all carriers. Other *Essex*-class carriers had different configurations. The catapults were not used much until later in the war as aircraft payloads became heavier. For landing aircraft, there were nine stern and six bow arresting wires. The *Essex* classes were the first carriers purposefully designed to launch or land aircraft from the stern or bow.

The hangar bay has been discussed in the flight deck and armor sections. In addition to the considerations mentioned above, the hangar bay was quite large and this facilitated storing and servicing a large number of aircraft. Since the ship class was required to carry upwards of a 25 percent supply of aircraft parts, the expectation was that large numbers of aircraft could be repaired and put back into service while at sea. The hangar bays were divided into three sections that were each configured differently. Firefighting equipment was improved but this will be discussed elsewhere. The hangar bays had large roll up doors on the sides that allowed for improved ventilation and the warming up of aircraft in the hangar bay but could still be kept closed in bad weather. This was also considered to be a firefighting advantage in that with the doors open and a sharp turn, burning debris, or fuel would hopefully wash overboard. All in all, the hangar bays were improved from the *Yorktown* class but not changed dramatically. The design requirement was to hold four squadrons of eighteen planes: one of fighters, one of scout dive bombers, one of dive bombers, and one of torpedo bombers with a second squadron of fighters in reserve. The scout bomber and the reserve squadron concepts were dispensed with early on, however, so they instead had two squadrons of fighters, two of dive bombers, one of torpedo bombers, and none in reserve. As the war progressed, this would change (more fighters which could also be used as fighter-bombers and less dive bombers being the main change).

The armor of the flight deck and hangar bays has already been discussed. Elsewhere, the ship was given a four inch belt of Class B armor (tapering to 2.5 inches at its lower edges). In many other parts of the ship, Special Treatment Steel (STS) was used as this was more resistant to damage and usable as construction material. Regular armor had to be placed over other metal but STS could be part of the frame and surface as well. The bulkheads were 4 inches of Class B armor as well. Other bulkheads had an additional 0.625 inches of STS-especially near the belt and hangar deck. The armor on the hangar bay was STS as well, along with many sections of the ship with the additional 0.625 inches of STS: magazines, aviation gasoline tanks, and other critical bulkheads. Other areas had even more, such as one inch on funnel uptakes, bridge, etc. There were also several dead spaces separated by multiple bulkheads that protected key areas. These could be filled with air, oil, water, etc. For protection against mines and torpedoes, a triple hull was installed. There was also extra armor along and below the waterline to try to protect from mines and torpedoes.

Overall, the decisions on armor plating were made to prevent the ship from being top heavy, to facilitate more area for aircraft storage and to protect the vital lower areas and island from splinters. As mentioned, this was a trade-off, and other countries did not feel the same. For example, most British carriers had armored flight decks but their aircraft complements were lower. The Japanese also tried armored decks later in the war, although they were too late to impact the fighting. Overall, the design incorporated everything known about ship armor and protection leading up to the war.

The power configuration was updated from previous classes of aircraft carriers. Eight Babcock & Wilcox boilers fed four steam turbines that in turn turned four shafts and four propellers. Each propeller was solid manganese bronze, 15 feet in diameter and weighed 27,000 pounds. This configuration generated 150,000 standard horsepower (SHP) and drove the ship at 32.5 knots. Their layout in the ship was better organized to try to prevent catastrophic damage to all in combat with each pair of boilers connected to a turbine but also cross-connected to supply the other turbines if needed. These boilers could provide 565 pounds per square inch (psi) each and a temperature of 850 degrees. This allowed these boilers to generate a higher power-to-weight ratio, making them more efficient than previous boilers. They were also lighter, used less fuel, and their uptakes were smaller. They were configured alternating between engine and fire rooms and this was supposed to prevent one hit from taking out all of the engines. They had a low-pressure and high-pressure turbine driving each propeller shaft (through a double-reduction gearbox) plus the option to use a cruising turbine for more economy. They also had to have high speed astern because, as noted before, they were expected to launch aircraft from the stern. For electrical power, the class relied on four 1,250-kilowatt turbo-generators. In case of damage, there were also two 250-kilowatt diesel generators as a backup. For low power needs, two ship's service generators were installed and these in turn were backed up by three 60-kilowatt emergency generators.

At 15 knots, the *Essex* class could travel 20,000 nautical miles (NM) without refueling. They could carry 6,330 tons of fuel oil as well as 240,000 gallons of aviation gasoline for the aircraft. The storage system for aviation gas was advanced. It had a cylindrical central storage area covered by two saddle-shaped storage areas that used the technique of pumping in sea water to provide pressure and prevent dangerous vapors from collecting. The water and gas were kept carefully separated and the whole tank was armored as well. Overall, the *Essex* class had a 67 percent improvement in range, 40 percent improvement in fuel storage, and 25 percent improvement in aviation gas storage compared to the *Yorktown* class.

For the *Essex* class's water needs, three distillation plants were installed. These boiled sea water to create fresh water from the vapor. High- and low-pressure air compressors were used to provide air for all the ship's needs. For firefighting, nine fire pumps were installed. Six bilge pumps were on hand to clear the bilges and pump out flooded areas.

For armament, the *Essex* class was similar to the *Yorktown* class again. However, during construction, multiple changes were made incorporating lessons learned from early war battles. The big lesson was that you could never really have too much anti-aircraft weaponry. For long-range anti-aircraft fire and for ship-to-ship combat should that ever prove necessary (it was not for the *Essex*-class carriers), each ship carried twelve 5-inch/38-caliber dual-purpose guns. Eight of these were in dual gun turrets, two each fore and aft of the island structure. Four more were in single gun mounts on the port side of the ship. Eventually they each held at least forty 40-mm Bofors anti-aircraft guns in ten quadruple mounts. Finally, they also had upwards of fifty-eight single 20-mm anti-aircraft guns. The 40-mms replaced the 1.1-inch anti-aircraft guns and the 20-mms replaced the 0.5-inch machine guns of the older carriers. The 5-inch guns were designed to intercept aircraft before they could start their attacks. The 40-mms were somewhere in between and the 20-mms were trying to destroy aircraft after they had released their payloads. Later, once *kamikazes* arrived, the 20-mms were critical in thwarting aircraft in the final parts of their dives against the carriers.

Safety improvements included having the hull divided into more watertight compartments (with more watertight bulkheads), having extra foam, fog, and salt water hoses on the various decks where needed, fire stations in each hangar bay (behind metal walls with thick glass to look through) that controlled the sprinkler systems and the magazines were floodable. These features were all supported by first class damage control training so that all the advanced features could be correctly utilized.

The island was made smaller and the deck guns were reconfigured to make more room for a larger flight deck. Two twin 5-inch gun mounts fore and aft of the island saved some space from the configuration on the *Yorktown* class. In addition, the third, folding, side elevator freed up space on the hangar bay and flight deck and still allowed the ship to safely navigate the Panama Canal.

Construction

The USS *Hornet* (CV-12) was built by Newport News Shipbuilding & Dry-dock Company (more commonly referred to as Newport News) in 1942 and 1943. Newport News has a long and robust history of shipbuilding for the US Navy that started when it was founded in 1886 as the Chesapeake Dry Dock & Construction Company. Built up at the end of a new railway that brought coal from West Virginia mines, they originally started by repairing ships that serviced this crucial transportation hub. The company officially opened when the USS *Puritan* (BM-1) docked there in 1889. Quickly they shifted into building ships

Shipyard workers preparing USS *Hornet* (CV-12) to slide down the ways, August 30, 1943.

for commercial and US Navy use. The first ship built there was the *Dorothy*, a tugboat in 1891. Soon after this, three gunboats were built for the US Navy: USS *Nashville* (PG-7), USS *Wilmington* (PG-8), and USS *Helena* (PG-9). All three ships played big roles in the Spanish-American War and beyond.

This was the start of a long relationship between Newport News (they changed their name in 1890) and the US Navy. They built six of the US Navy's early dreadnoughts from 1906 to 1923, twenty-five destroyers from 1918–1920, and more importantly for the USS *Hornet* (CV-12), they began building aircraft carriers in the 1930s. They launched the USS *Ranger* (CV-4) in 1933 and the USS *Yorktown* (CV-5) and USS *Enterprise* (CV-6) in 1936. This was important as they had experience building modern, purpose-built aircraft carriers. Eventually Newport News Shipbuilding would construct the USS *Essex* (CV-9), USS *Yorktown* (CV-10), USS *Intrepid* (CV-11), USS *Hornet* (CV-12), USS *Franklin* (CV-13), USS *Ticonderoga* (CV-14), USS *Randolph* (CV-15), USS *Boxer* (CV-21), and USS *Leyte* (CV-32) of the *Essex* class. Overall Newport News would be rated E for excellent by the US Navy for their shipbuilding in World War II that included a total of 243 ships (186 of which were *Liberty* ships).

The contract for the USS *Hornet* (CV-12) was awarded to Newport News on September 9, 1940. Eventually the keel was laid on August 3, 1942, using slipway #8, which had built the USS *Hornet* (CV-8) and the USS *Essex* (CV-9). This ship was originally designated as the USS *Kearsarge* (CV-12), but when the USS *Hornet* (CV-8) was sunk in the Battle of the Santa Cruz Islands on October 26, 1942, the navy decided to change the name to the USS *Hornet*. Deep within the hull of the USS *Hornet* (CV-12), you can still find the name USS *Kearsarge* stamped into the metal.

Laying the keel consists of building a large steel "I" beam built out of flat steel plates set as angles to form the spine of the ship essentially. In the case of the USS *Hornet* (CV-12), this was 870 feet long. From here the ship was built up quickly with more than 2,000 workers working in shifts that covered all seven days of each week. They added all the rest of the framework and then each compartment. There were 9,160 different plans that guided the workers as they constructed the ship. Even with all the plans, changes were made during construction as technology changed and war lessons were incorporated. This mostly resulted in improvements to the radar systems and the anti-aircraft gun configuration as noted before.

Like every other *Essex*-class carrier built, the USS *Hornet* (CV-12) was completed ahead of schedule at a cost of $69 million. This was due to several factors in the design of the *Essex* class but most importantly the high number of large, flat steel plates used as well as the decision to primarily use welding instead of riveting. The ship was launched on August 30, 1943 only one year and three weeks after the keel was laid. The ship's sponsor was Annie Reid Knox, who just happened to be the wife of Secretary of the Navy Frank Knox. After a short delay due to issues with the hydraulics used to launch the ship, the ceremony was held with Mrs. Knox breaking the champagne bottle against the hull and the USS *Hornet* (CV-12) slipping into the water of the James River.

Above: Shipyard workers preparing USS *Hornet* (CV-12) to slide down the ways (2), August 30, 1943.

Right: USS *Hornet* (CV-12) starts to slide down the ways into the water, August 30, 1943.

Above: Edward B. Harp, CDR USN, chaplain of USS *Hornet* (CV-8), aboard when it was sunk, prepares Champagne bottle to christen USS *Hornet* (CV-12), August 30, 1943.

Opposite: USS *Hornet* (CV-12) slips into the water with Secretary of the Navy Knox and guests in attendance, August 30, 1943.

Of course, the launching of the ship does not complete its construction. The slipway needed to be freed up for the next ship to be started. Around three months of work still remained, but the rest of this could be done while the ship was tied up to a pier. When this process started, the ship was roughly 70 percent complete. The island was added along with the masts, machinery, guns, and much of the electrical, plumbing, and hydraulic systems. The ship also had to be made ready for its crew that started arriving. This included finishing the living quarters, galleys, machine shops, engines, boilers, and medical facilities. In addition, as the crew began to arrive the ship had to start loading on tons of supplies, from food to ammunition and fuel. The originally planned crew was 230 officers and 2,256 chief petty officers and enlisted men. This number would grow throughout the war as more anti-aircraft guns were added and other changes made, eventually reaching 3,400 total crewmen. Finally, after around the clock work, the ship was commissioned on November 29, 1943 and was almost ready to join the US Navy. Captain Miles Rutherford Browning had been named as her first commander and he was given the battle flag from USS *Hornet* (CV-8). By December 20, 1943, USS *Hornet* (CV-12) was officially ready to begin her sea trials.

Secretary of the Navy Frank Knox leans on the USS *Hornet* (CV-12) while his wife, Mrs. Knox, christens the ship. RADM O. L. is to the left of Mrs. Knox, August 30, 1943.

Specifications

Displacement:	27,100 tons standard; 36,380 tons full.
Length:	872 feet (short hull variant including USS *Hornet*), 888 feet (long hull variant).
Beam:	147 feet (waterline: 93 feet).
Draft:	29 feet.
Machinery:	Eight Babcock & Wilcox boilers, four Westinghouse geared steam turbines, four shafts/four propellers.
Bunkerage:	6,330 tons fuel oil, 240,000 gallons of aviation fuel.
Power Output:	150,000 SHP.
Speed:	33 knots.
Range:	20,000 nautical miles (nm) at 15 knots.
Crew:	2,630–3,448 (the number increased over time as changes were made).
Armament:	Twelve 5-inch/38-caliber guns, Thirty-two to seventy-two Bofors 40-mm 56-caliber guns, fifty-five to seventy-six Oerlikon 20-mm 78-caliber guns.
Armor:	2.5-inch to 4-inch belt, 1.5-inch hangar and protective decks, 4-inch bulkheads, 1.5-inch STS top and sides of pilot house.
Aircraft:	Ninety to 100.
Elevators:	Three (two centerline, one deck-edge).
Radar:	One SK air-search radar, one SC air-search radar, two SG surface-search radars, two Mk. 4 fire control radars, ten to seventeen Mk. 51 AA directors.

2

World War II
Part 1

Preparing for Battle

Captain Miles R. Browning was destined to be a controversial and short-lived commander of the *Hornet*. In some ways, he had a very impressive record and was by all accounts a brilliant tactician. However, he was very difficult to get along with, was a heavy drinker, and had a knack for making enemies of the wrong people.

Browning graduated from Annapolis as a commissioned ensign on June 29, 1917. He served on multiple ships until reporting to Pensacola in January 1924 for flight training. He ended up being one of the early navy combat pilots on the *Langley*. He was considered a skilled pilot, but even back then was considered somewhat out of control. In 1931, he moved to the bureau of aeronautics and served as a test pilot and aircraft designer. He later took command of a fighter squadron on the USS *Langley* (CV-1) and then the USS *Ranger* (CV-4). His impressive career continued with post-graduate work at the Navy War College, serving as a pilot trainer at the Air Corps' tactical school and then finally becoming a member of Admiral Halsey's staff in June 1938. By 1941, Browning was Admiral Halsey's chief of staff, and after Pearl Harbor, it was his planning and tactics that led to the successful raids by the USS *Enterprise* (CV-6) and other US carriers. He helped to plan the Doolittle Raid and when Halsey was too sick for Midway, Browning stayed on to support Admiral Spruance in that stunning US victory. Browning is credited with wanting to time the US strikes to hit the Japanese carriers while they were refueling and rearming their aircraft. Spruance found Browning very difficult to get along with (like others did), and despite his tactical brilliance, he was always on the edge of serious trouble. Halsey favored and protected him however and after reasonably favorable results in the Guadalcanal campaign, Browning was rewarded with a ship of his own, the *Hornet*.

The USS *Hornet* (CV-12) was finally commissioned on November 29, 1943 and left Norfolk on December 20, 1943 for two weeks of sea trials on Chesapeake

USS *Hornet* (CV-12) leaving Pearl Harbor on the way to her first combat mission, March 15, 1944.

Bay. Sea trials were used on new ships by the navy to start the process of training and familiarizing the crew as well as testing out its capabilities. The crew ran trials of the speed, endurance, maneuverability, seaworthiness, and fuel consumption of their ship. They also had practice firing drills and other training exercises. Issues were noted for fixing when back in port.

Air Group Fifteen joined the ship with their thirty-six F6F-3 Hellcats (VF-15), thirty-two SB2C Helldiver dive bombers (VB-15), and sixteen TBF-1 Avengers (VT-15). Limited air operations were started and the first arrestor gear landing on the *Hornet* was made on January 1, 1944. These pilots had been carrier qualified elsewhere but needed to familiarize themselves with the unique configuration of their new *Essex*-class carrier.

After a brief return to Norfolk for topping up of provisions and an inspection by the Norfolk Fleet Air Commander (and staff), the *Hornet* and her crew were underway on January 15, 1944 bound for the Eastern Sea Frontier off of Bermuda for their shakedown cruise. The *Hornet* was accompanied by the destroyers USS *Forrest* (DD-461), USS *Corey* (DD-463), and the USS *Hobson* (DD-464). The *Hobson* was later replaced by the USS *Carmick* (DD-493). Normally a shakedown cruise was supposed to last four to five weeks but the war in the Pacific was not waiting for the *Hornet*, and against Captain Browning's wishes, he was only given two weeks.

Off of Bermuda starting on January 17, 1944, the *Hornet* practiced ship maneuvers, radio and radar use, anti-aircraft fire (sometimes against towed targets), refueling at sea (and fueling destroyers from the *Hornet* at sea), and all of the other

VT-15 Avenger in flight, February 22, 1944.

hundreds of activities the ship would need to perform in actual combat. All types of ship's guns were fired, all radios were calibrated, all types of crew drills were run, and all of the different ship's speeds and maneuvers were practiced. For the air group, the training included 1,225 aircraft landings and thirty-five catapult launches. They practiced bombing, fighter interceptions, and more. They were unable to practice torpedo attacks except as simulation. The pilots and aircrew logged over 2,000 hours in the air. The training was realistic enough that six aircraft were lost to accidents along with the lives of three airmen during this time. Both incidents that led to loss of life were in Helldivers: the first was a spinout where the radioman was rescued but not the pilot and the second was when a plane failed to pull out of a dive, losing both crewmen. This shortened shakedown cruise was to have large ramifications when the *Hornet* eventually reached Pearl Harbor.

After a final return to Norfolk on February 1, 1944, the *Hornet* stayed in dock for two weeks as every provision was fully stocked, every piece of machinery was checked and double-checked and the crew said goodbye to Virginia. Aviation gas and fuel oil were topped off, wrecked and damaged planes were removed and replaced by new ones, and repairs were conducted on items noted during the shakedown cruise. These repairs included the SN radar and several proposed fixes to aircraft. In particular, the Avengers and Helldivers had multiple issues that were worked out by the time the ship left.

On February 14, 1944, the *Hornet* left Norfolk to head for the Panama Canal on the way to San Diego and the Pacific. The first destination was Puerto de

Simulated torpedo boat attack on USS *Hornet* (CV-12) near Panama Canal, February 24, 1944.

Cristobal, on the Caribbean side of the Panama Canal in the company of USS *Albert W. Grant* (DD 649) and USS *J. D. Blackwood* (DE 219). During this cruise, the crew continued to practiced gunnery and air exercises. This helped to partially rectify the short shakedown cruise time. On February 19, the *Hornet* passed through the Panama Canal and transferred to the control of commander in chief of the Pacific Fleet. The *Hornet* was assigned as the leader of task group 12.1 in the company of the USS *Albert W Grant* (DD-649) and the USS *Melvin* (DD-680) for the trip to San Diego.

On the way to San Diego, the crew continued to practice. Gunnery drills included radar guided firing of the 5-inch guns, refueling destroyers at sea and continuous air operations, including such drills as repelling torpedo plane attacks. Arriving in San Diego on February 27, 1944, the *Hornet* restocked all supplies, exchanged damaged aircraft for new ones, and embarked 2,000 marines and some of their equipment to take to Pearl Harbor. On February 29, the *Hornet* left San Diego for Pearl Harbor and their first real assignment in the war zone. The crew continued to train and practice as they crossed the Pacific, finally arriving on March 4. The marines and their equipment were offloaded and the *Hornet* prepared to enter the battle by again topping up on supplies and replacing damaged aircraft.

While in Pearl Harbor, the *Hornet* started training that went on for over a week with USS *Morris* (DD-417), USS *Mustin* (DD-413), and USS *Hughes* (DD-410) as escorts. Captain Browning made a controversial decision just before the training started. He deemed that Air Group Fifteen not to be ready for combat and asked

VF-15 Hellcat attached to hangar bay catapult on USS *Hornet* (CV-12), February 25, 1944.

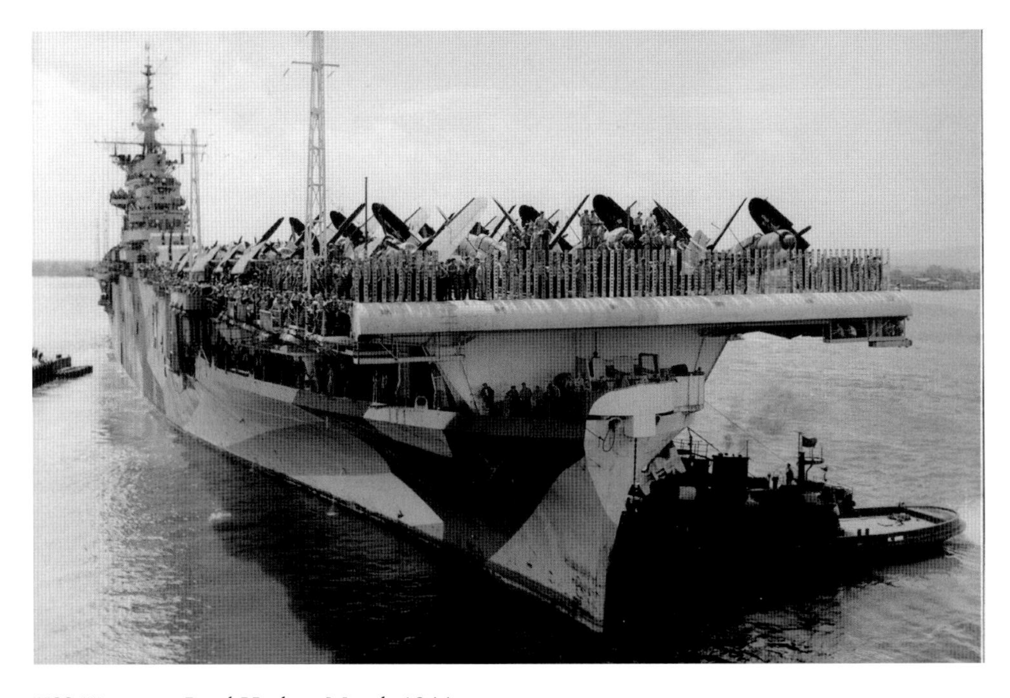

USS *Hornet* at Pearl Harbor, March 1944.

for them to be replaced with Air Group Two that was at Pearl Harbor. This request was granted and Air Group Fifteen left the ship and was replaced with Air Group Two on March 8, 1944. All the camaraderie that had been built between the sailors and airmen of the *Hornet* had to start over with the new air group. The new air group was trained and requalified on carrier landings.

Air Group Two had a long and impressive record. The original Air Group Two was well known through the navy in the 1930s. In July 1937, they participated in the unsuccessful search for Amelia Earhart. In 1938, they participated in Fleet Problem XIX and were successful in delivering a sneak attack on Pearl Harbor. Once World War II started, they were assigned to the USS *Lexington* (CV-2) and participated in several early war battles. This included a successful attack over the Owen Stanley Mountains in New Guinea to strike Japanese ships that were part of an attack there. Several transports were sunk in this operation (one of the first really successful attacks by the US Navy in World War II). Shortly after this was the Battle of the Coral Sea. The US Navy sank the Japanese light carrier *Shōhō* but lost the USS *Lexington* (CV-2) and suffered heavy damage to the USS *Yorktown* (CV-5) during the battle. With their carrier gone and many aircraft lost, Air Group Two was disbanded. However, a new Air Group Two was formed in 1943 and was based at Quonset Pt. Naval Air Station (NAS) in Rhode Island. Designated as a night group, they trained extensively at night until reclassified as a "regular" air group. Eventually all the elements of the air group were transported to Pearl Harbor in late 1943 on small and jeep carriers. VF-2 spent some time on the USS *Lexington* (CV-10) and USS *Enterprise* (CV-6) in late 1943. All elements of the air group were reunited in Hawaii until assigned to the USS *Hornet* (CV-12) in March 1944.

Finally on March 15, 1944, the *Hornet* left Pearl Harbor to officially go to war. On board was Rear Admiral J. J. Clark (commander of Task Force 58.3), who was not a fan of Captain Browning. The *Hornet* traveled with USS *Cowpens* (CVL-25), USS *Monterey* (CVL-26), USS *Izard* (DD-589), USS *Charette* (DD-581), USS *Connor* (DD-582), USS *Bell* (DD-587), USS *Burns* (DD-588), USS *Caldwell* (DD 605), USS *Bradford* (DD-545), USS *Cowell* (DD-547), USS *New Orleans* (CA-32), and USS *Indianapolis* (CA-35). The *Hornet* steamed to Majuro Atoll, then an advanced base of the US Navy, arriving on March 20 and the ships were reconfigured into different task forces, with *Hornet* joining Task Force 58.2 under Rear Admiral A. E. Montgomery. As usual intensive training continued while *en route*. Task Force 58.2 aircraft carriers included *Essex*-class carrier USS *Bunker Hill* (CV-17), *Independence*-class carriers USS *Monterey* (CVL-26) and USS *Cabot* (CVL-28), and now USS *Hornet*. The whole Task Force 58 was under the command of Admiral Mistcher. The rest of Task Force 58 included the fleet carriers USS *Enterprise* (CV-6), USS *Yorktown* (CV-10), USS *Lexington* (CV-16), USS *Princeton* (CVL-23), USS *Langley* (CVL-27), USS *Cowpens* (CVL-25), plus supporting cruisers, destroyers, and battleships. The US Navy had come a long way from the early part of the war.

What was the situation that the *Hornet* was entering in early 1944? It was very different from the dark days of early 1942 and the long, grinding attrition of

Curtiss SB2C Helldiver crashes on the flight deck of USS *Hornet* (CV-12) and catches fire. Crew fights fire, March 7, 1944.

1943. After their spectacular victory at Pearl Harbor on December 7, 1941, the Japanese had swept across the Pacific in only three months, conquering virtually all of their objectives: Wake, Guam, the Philippines (even though some US troops still held out until May 1942), Malaysia (including Singapore), Burma, the Dutch East Indies, the north and east coast of New Guinea, and the Solomon Islands. Despite a few, minor allied successes, ship, aircraft and troop losses had been very low for the Japanese up to that point.

The US and Allied navies mostly had to content themselves with raids and holding actions. Since the US carriers had not been found at Pearl Harbor, the US Navy was able to conduct a series of raids on Japanese bases across the wide Pacific Ocean, with the most famous being the Doolittle Raid in April 1942. By bombing the Japanese mainland, the US proved that it was not ready to give up. These raids were not going to be enough to turn back the Japanese, although they did provide excellent training opportunities for the US carriers.

The Japanese wanted to isolate Australia from the US. To accomplish this, they wanted to capture Port Moresby on the southern coast of New Guinea and they wanted to work their way down the Solomon Islands to Guadalcanal where they would build an air base to threaten the shipping lanes between Australia and the

B-25 Bombers on the deck of USS *Hornet* (CV-8) for the Doolittle Raid, May 1942.

US and possibly even capture New Caledonia and Fiji. The US was going to have to start fighting directly if they wanted to prevent this from happening.

The Battle of the Coral Sea was the first such major clash. The Japanese were attempting to capture Port Moresby and they had two of their fleet carriers, the *Shōkaku* and the *Zuikaku* along with a light carrier *Shōhō* to escort their invasion fleet and fight off any US interference. The US countered with the USS *Lexington* (CV-2) and the USS *Yorktown* (CV-5) and the first battle ever between two fleets that never saw each other ensued. In a confusing, chaotic, multi-day battle, the Japanese won a tactical victory by trading the giant USS *Lexington* (CV-2) for the tiny *Shōhō*. However, the strategic victory went to the Allies as the invasion fleet was turned back and Port Moresby was saved. Even more importantly, the *Shōkaku* was so damaged and the air groups of both Japanese carriers were so depleted that neither participated in the coming Battle of Midway. For the US, the USS *Yorktown* (CV-5) was also heavily damaged but was repaired in just three days at Pearl Harbor and did participate in the Battle of Midway.

The Battle of Midway was the Japanese grand plan to draw the rest of the US Fleet and especially the carriers into battle so they could be destroyed once and for all. Most of the Japanese Navy participated in this battle, but the complicated battle plan had their ships scattered into separate fleets. This ensured that the main battle was between carrier fleets with four Japanese carriers (*Hiryū, Sōryū, Akagi*, and *Kaga*) and three US carriers (USS *Enterprise* [CV-6], USS *Hornet* [CV-8], and USS *Yorktown* [CV-5]). By benefit of breaking the Japanese naval codes, the US knew where the Japanese were going to strike and they laid a trap for them. The plan worked, with a healthy dose of luck, and when the battle ended the Japanese invasion was called off and they lost all four of their carriers (and a heavy cruiser) while the US only lost the USS *Yorktown* (CV-5) and a destroyer. The US Navy had managed to win the most significant victory in its entire history.

After this, a battle of attrition started between the US and Japanese navies and air forces in the south pacific that lasted well into 1943. This started when US Marines landed on Guadalcanal in August 1942 and captured the Japanese airstrip there. This coincided with counterattacking US and Australian troops on New Guinea. Over the next year plus, this resulted in almost non-stop air, sea and land conflict between the Japanese and the Allies. While the Japanese had several tactical victories (especially in night surface combat where their superior training and torpedoes more than compensated for Allied radar), the strategic situation tilted inexorably towards the Allies. The US carrier forces were whittled down with the USS *Wasp* (CV-7) and USS *Hornet* (CV-8) being lost in exchange for two light Japanese carriers, the *Ryūjō* and the *Chūyō*. However, many other Japanese carriers were damaged or had their air groups crippled during this time. But this did not matter as long as the US held "unsinkable" Guadalcanal and air bases on New Guinea. With US Navy, Marine and Army Air Force planes operating from these areas, a steady attrition of pilots, planes, and ships was inflicted on the Japanese. Not only were their losses heavier but their ability to replace those losses was much lower. US and Allied F4F Wildcats, SBD Dauntless dive bombers, P-39 Airacobras, P-38 Lightnings, B-25 Mitchells, and B-17 Flying Fortresses held their own and steadily wore down their opponents.

Because of heavy carrier and aircraft losses, both sides spent 1943 rebuilding their carrier forces. The Japanese managed to launch some new carriers as well as complete some ship conversions into carriers. For the US, the *Independence*-class light carriers and the *Essex*-class carriers were swinging into full production and starting to come online in quick succession. One major difference was that combat experienced Allied pilots were rotated back to the mainland to help train the next generation of pilots. On the Japanese side, pilots stayed on the front lines until lost. This meant that the heavy losses in the south pacific slowly killed off most of their experienced pilots. The new replacement pilots had much less training than their predecessors and did not benefit from the combat experience of the better pilots that were left. This was both because of fuel shortages and because losses had been so heavy that pilots had to be moved to the front faster.

In the area of submarine warfare, the US Navy was rapidly eclipsing their Japanese counterparts. With their torpedo problems sorted out, new submarines joining the fleet regularly and with the "front line" of the Pacific being pushed back allowing for more advanced bases, US submarines made their mark on the Pacific War. They ventured far and wide, scouting Japanese fleet movements and decimating their merchant fleet. While Japanese submarines had notable successes early in the war (finishing off the USS *Yorktown* [CV-5], sinking the USS *Wasp* [CV-7], and the USS *Liscome Bay* [CVE-56]), these were not happening anymore.

On top of this, the most significant difference between the Japanese Navy and the American Navy was the quality of aircraft. For the most part, at the beginning of the war, the Japanese planes were superior. They were more maneuverable, had better range, better climbing rates, etc. and, most importantly, had excellently trained and combat experienced pilots. Once some of their weaknesses were discovered (lack of self-sealing fuel tanks and armor most notably), and new tactics were invented, the early Allied aircraft found that they could hold their own. The classic match up between the US Wildcat and the Japanese Zero was more or less a toss up eventually (this was helped by the fact that the US forces were often fighting defensively so even if their planes were shot down the pilots could often be rescued). However, as the conflict moved into 1944, the Japanese Navy had not upgraded their aircraft as much as the Allies. They had replaced their Val dive bombers and Kate torpedo bombers with Judys and Jills respectively and the Jill was especially a big improvement. The US Navy had upgraded from their Devastator torpedo bombers and Dauntless dive bombers to the Avenger and Helldiver, both better aircraft. The Japanese Navy had a big deficiency in fighter planes, however, as they were still primarily using their only marginally updated Zeroes (also referred to as Zekes). In the meantime, the US Navy had upgraded from the Wildcat to the Hellcat. The Hellcat was superior in just about every way except range and turning. It was well armored, climbed quicker, flew higher and faster, was well armed and benefited from having extremely well-trained pilots. In addition, the Marines were now using the F4U Corsair, another plane superior to the Zero in almost every way. The difference in fighter aircraft and pilot training/experience was to be the most significant factor in the naval battles fought for the rest of the war.

As far as the battle for the bases and islands of the Pacific was going, the Japanese were fully on the defensive. In the South Pacific, the Allies had pushed the Japanese back in their campaign in New Guinea, slowly driving the Japanese from the east coast and then onto the northern coast. At the same time, the Allies had moved up the Solomons islands, capturing or neutralizing Japanese bases and encircling Rabaul, the main Japanese stronghold in the area. In the Central Pacific, the Allies had captured bases in the Gilberts and Marshall islands, driving a wedge into the Japanese perimeter. None of these actions had generated large responses from the rebuilt Japanese carrier forces however. For the most part, the battles were over too quickly or were within range of land based air, making a sortie by the Japanese fleet too much of a risk for too small of a gain. Several large US carrier raids had demonstrated the effectiveness of fast carrier forces, including a devastating raid on Truk, Japan's premiere naval base in the Pacific.

The next steps in driving on to the northern coast of New Guinea and into the Marianas were sure to generate a response from the Japanese. These attacks brought the US too close to Japanese shipping lanes from the East Indies on the one side and within range of the Japanese homeland for B-29 Superfortresses on the other. In addition, both lines of attacks brought the Allies closer to the Philippines/Formosa/Bonin Islands and from any of these locations, the Allies would have advanced bases far too close to the Japanese home islands. The vastness of the Pacific Ocean served a similar fashion to the vastness of Russia, giving the defender time to fall back before the attacker could overwhelm them.

However, the Japanese were running out of places to fall back to. The Japanese had to respond to these next attacks. A showdown was coming after almost a year without major carrier battles in the Pacific.

This was the situation the *Hornet* and her crew found themselves in when they joined Task Force 58 in March 1944: a large and growing US Navy that now had superior equipment in almost every way facing a weakening Japanese Navy that still had a lot of fight left and needed a decisive victory to turn the war around before they were overwhelmed by US industrial might. Despite a "Germany First" strategy, the Allies were able to aggressively push the Japanese back on two fronts, in the southwest Pacific and across the central Pacific. The Japanese were not without hope, however, as a large portion of their surface fleet was still intact, including the super-battleships *Yamato* and *Musashi* and their carrier force had been regrown after the losses at Midway and in the Solomons. They could still hope to redress the balance of forces with a dramatic victory against US naval forces. This was the idea of Operation Z, a plan put together by Admiral Mineichi Koga in response to Allied advances. Their basic plan was to wait until the US was conducting a large amphibious operation that would tie the fleet in place for some time and then attack it with land-based air units and the largest portion of the remaining Japanese fleet that could be assembled. This would create the long

USS *Hornet* (CV-12) damage repair crew fight fire of burning SB2C Helldiver (VB-2), April 21, 1944.

hoped for *Kantai Kessen*, the single dramatic battle that would decide the war. The Japanese needed to do this while the overall numbers of ships and aircraft was not too unbalanced against them. After all, this is what the US had done at the Battle of Midway in 1942 (the last time the Japanese seriously tried to invoke the *Kantai Kessen*). If they could win the air battle, they might be able to engage in a large surface battle that they could still hope to emerge victorious from. The Japanese still hoped that a significant setback for the US could bring them to the bargaining table.

The *Hornet*'s First Carrier Raids

The *Hornet* and her crew were finally going to enter combat for the first time in late March 1944. Task Force 58 left Majuro on March 22 to undertake Operation Desecrate One, a large raid on Japanese held islands in the central Pacific in support of amphibious operations scheduled in New Guinea. The first part of this was an air raid on the Palau Islands. The Palau Islands held important harbors and air bases and were located almost like a shield in front of the Philippines. Ships and aircraft from these bases could interfere with operations in New Guinea and elsewhere. This was the first raid against the Palau Islands in World War II, although the Allies would launch a full invasion of Peleliu in September 1944. The Japanese spotted the US task force on March 25 and this gave them enough time to retreat some of their naval units. This action was not without cost to the Japanese as it led to the death of Admiral Koga, Yamamoto's successor. The Japanese knew they did not have enough air units in the area to protect their ships.

On March 28, 1944, an Avenger from *Hornet* spotted a G4M "Betty" bomber and chased it but was unable to catch it. Now Task Force 58 knew it had been spotted. On March 29, fighters from the *Hornet* scored their first victory as they splashed a Betty being used to scout the US fleet. That night the Japanese tried to make torpedo attacks on the US fleet under the cover of darkness. A combination of night fighters and anti-aircraft gunnery drove them off.

The attack started on March 30, 1944 with a fighter sweep over the airfields of the Palaus in which thirty Japanese fighters were shot down, including nine claimed by *Hornet* pilots. Following this, Avenger bombers came in and dropped mines around the harbors and sea channels of the area to prevent Japanese ships from escaping. This was the first use of this tactic and proved relatively successful. Over March 30 and 31, the carrier planes shot down or destroyed on the ground virtually all of the Japanese aircraft in the area (114 in the air and forty-six on the ground); in addition, they heavily damaged harbor and ground facilities and sank or damaged thirty-six ships. Among the ships sunk were a destroyer, a torpedo boat tender, a submarine tender, an aircraft transport, a repair ship, and three tankers. All in all a great start for the *Hornet* as part of Task Force 58.

Task Force 58 turned back east after this attack but was not done fighting yet. Raids on several more island groups commenced: Ulithi, Yap, Woleai, and the

SB2C Helldivers (VB-2) parked on the flight deck of the USS *Hornet* (CV-12), April 28, 1944.

Damaged SB2C Helldiver (VB-2) on flight deck of USS *Hornet* (CV-12), April 21, 1944.

Caroline Islands. These raids went as well as the attack on the Palaus, although there were significantly less ships to attack. Japanese air defenses were swept away and installations and the few ships found were destroyed. By April 6, 1944, Task Force 58 had returned to Majuro atoll for a week's rest after a job well done. No damage was suffered by any of the US ships and aircraft losses were light. Twenty-five planes were lost in combat and another eighteen to operational accidents. While forty-four aircrew were shot down, twenty-six were rescued (the US Navy had successfully started using submarines stationed near targets to recover airmen shot down). The Japanese simply did not have an answer for these raids and could only conserve their forces until the US committed to a large, lengthy invasion somewhere. Observations by the US Navy during this battle were that fighter sweeps, mine-laying by torpedo bombers and delayed fuses for their bombs were all very successful. Dive bombing needed to improve and more emphasis was needed on "cratering" airfields because of the Japanese tendency to fly in aircraft reinforcements at night.

The following is a detailed listing of the air activities specific to the *Hornet* during this battle. This is given by way of an example of what carrier operations were like. Normally there were also CAP fighters and anti-submarine flights but for this mission those duties were handled by the CVLs to allow the big carriers to send out coordinated strikes. Finally, during combat aircraft from other carriers will sometimes have to land on different ones if they are damaged or short on fuel-another complexity. March 30, 1944 was a busy day for the *Hornet* as it was for all the carriers:

Mining group 1 launched with eight Hellcats and six Avengers at 6:30 a.m. to mine the approaches to Koror harbor.
Strike 1A launched with sixteen Hellcats, twelve Helldivers, and twelve Avengers also at 6:30 a.m. to hit shipping in Palau Harbor and W. Lagoon. Two Avengers were lost to operational accidents after take-off.
Strike 1B was launched with nine Hellcats and fifteen Helldivers at 8:45 a.m. against shipping again.
Strike 1C was launched with eleven fighters and nine Helldivers at 11 a.m. to attack the airfield on Peleliu Island.
Strike 1D was launched with eleven fighters and nine Helldivers at 1 p.m. to attack shipping again.
Mining Group 2 was launched with three Hellcats and four Avengers at 3 p.m., this time to mine Denges passage.
Strike 1F was launched with six Hellcats and four Avengers to attack anti-aircraft batteries and airfields.
Various photographic missions were launched during the day as well, to photograph damage and seek targets that might have been missed.

March 31 was not as busy for the *Hornet's* air group:

Strike 1A launched with twelve Hellcats, eleven Helldivers and six Avengers at 7:30 a.m. to attack shipping and drop more mines.
Strike 1B launched with twelve Hellcats and five Avengers at 9:30 a.m. One Avenger crashed after take-off.
Strike 1C launched with five Helldivers at 10 a.m.

April 1 was the final day of this operation and the *Hornet* was busy again:

Strike 1A launched with twelve Hellcats, eleven Helldivers and five Avengers at 6:30 a.m. against shipping and airfields in Woleai.
Strike 1B launched with twelve Hellcats and eleven Helldivers at 9 a.m. against the same targets. One Helldiver crashed after launch.
Strike 1C was launched with twelve Hellcats, eight Helldivers and six Avengers at 11 a.m. with the same mission.

After a week resting and replenishing in Majuro Atoll, Task Force 58 left on April 13, 1944 in support of the landings in Hollandia and Ataipe in New Guinea. These two operations were called Reckless and Persecution respectively. The operation was to be conducted in the same manner as the attack on the Palaus Islands. This time the carrier strikes were not really needed as the Army Air Force had done such a good job of attacking Japanese air bases in the islands before the carriers showed up. They were able to surprise the Japanese who thought they were out of range of fighter escorted US bombers. The Army Air Force had modified their P-38s so they had the range to accompany the bombers all the way to Hollandia. The *Hornet*'s aircraft bombed and strafed with all the other carrier aircraft starting on April 21 and provided close support to the landings. Normally carrier aircraft were used where land-based aircraft could not reach, but in this case both were used. No one expected the Army Air Force attacks to be as successful as they were. The *Hornet*'s task force struck Wadke, Sawar, and Sarmi. Instead of returning directly to Majuro atoll, Task Force 58 stopped off at the new base in Manus in the Admiralty Islands. Again, damage to Task Force 58 was negligible with losses of only twenty-one planes and twelve air crew.

After the devastating carrier raid on Truk in February 1944, it was hoped that it was finished as a base for the Japanese. An Army Air Force daylight bombing raid on Truk on March 29, 1944 encountered a large number of Japanese fighters, demonstrating that the Japanese had reinforced the base. Admiral Nimitz had wanted to order a follow-up carrier strike on Truk and this information confirmed his decision. Truk's location still made it dangerous have as an operational base so it had to be attacked again. Task Force 58 was assigned to return for another strike, this time with the *Hornet* as part of the attack force. Task Force 58 sailed north from the Admiralty Islands and on April 29 commenced the attack on Truk with a fighter sweep as was becoming standard practice. Japanese fighters rose to contest this attack and later torpedo planes were sent to attack the US fleet. By the end of the next day, at least ninety Japanese aircraft had been destroyed in

Burial at sea of John Joseph Chmura (VT-25, USS *Cowpens* [CVL-25]) who died aboard USS *Hornet* (CV-12) from injuries sustained while in action over Moen Island, May 1, 1944.

the air and on the ground, and almost no damage had been done to US ships (the *Lexington* suffered a near miss). The quality of the Japanese pilots was notably lower. The US lost twenty-six aircraft but all save nineteen air crew were rescued, demonstrating again the success of US submarine, destroyer and seaplane rescue of downed pilots. US surface ships were sent in to shell the bases in and around Truk. At this point, the Japanese gave up on using Truk as a base. Task Force 58 returned to Majuro atoll on May 4, 1944. They were to spend the next month preparing for the invasion of the Saipan in the Marianas.

The pilots and crew of the *Hornet* were learning their trade in combat situations and by all measures were excelling. Even though the ship had suffered no direct damage from Japanese aircraft, accidents were always a problem. Helldiver pilot Lt. Jack Taylor had a 100-pound bomb that would not jettison and so he was told to land with it. As soon as he landed, the bomb broke loose, bounced across the deck, and exploded, killing two sailors. Thirteen others were injured.

Gene Millen, fire controlman 3rd class, who served on the *Hornet* throughout its time in World War II, recalled many such incidents. He was in charge of ensuring the gun sights on the anti-aircraft guns were always functional and he found himself up near the flight deck watching air operations a lot. During an interview he gave in 2021, he talked about some of the things he witnessed:

> One time I saw a sad thing. One of our dive-bombers had a 100 pound bomb under each wing. One of them was hanging by, there was [*sic*] two hooks. And it was

A 100-lb. bomb dislodged from Helldiver during landing, killing two and injuring thirteen on USS *Hornet* (CV-12), May 1, 1944.

> hanging from one hook, and the bomb was off and when the plane landed the bomb exploded and the only injury was the guy that came out to unhook it. Another crash destroyed five gun sights and that was the busiest day of my life because I had to replace them. We carried spare parts of course. The only thing is the spare parts were on the sixth platform. All the hatches had to be closed. I got down there to the sixth deck. Each deck I had to go through the scuttle and close it, open the next one, go down, close it. All the way down and then I had to bring each gun sight up one at a time.
>
> Another time a pilot didn't turn the safety switch off and when he landed all six machine guns were firing. Fortunately nobody got shot because he was high enough.
>
> [Another incident] I remember now, the plane was one of the anti-submarine ships [*sic*], the Avenger, circling around the ship all the time and looking for stuff. Something happened with the plane and they had to make a forced landing and the plane lands in the water. They were in the water, waving, smiles on their faces. And the depth charge went off. That's the last you saw of them.
>
> [I saw a] twin engine Betty flying towards superstructure. Could see tracers from 20-mms going through aircraft and they stopped it and it took a nose dive and didn't hit our ship.

Millen saw a Helldiver successfully land with a hole in one wing that "four guys could stand in." All of this illustrates that even when not menaced with enemy ships and aircraft, naval aviation is a dangerous business.

During this time before the invasion of the Marianas, the unpopular Captain Browning was found to be in dereliction of duty and relieved of command. On May 15, 1944, while many crew members were watching a movie in the hangar bay, a fire extinguisher accidentally went off and they panicked, thinking it was a bomb. In the confusion there were some minor injuries and a crewman was knocked overboard. When rescued, he stated that he was pretty sure another sailor had fallen overboard. The crew was preparing a boat to find the sailor but they were stopped by Captain Browning and not allowed to do a crew count to see if someone was missing. This was even after Admiral Clark suggested he do both. A few days later, the body of the other sailor, who had indeed fallen overboard, was found. Admiral Clark convened a board of inquiry and Captain Browning was found guilty and removed from command. He spent the rest of the war commanding a naval air station.

Captain William Dodge Sample took over command of the *Hornet* and morale immediately improved. Captain Sample was an experienced commander and aviator, having served on a variety of cruisers and battleships in their aviation departments as well as on several carriers. He also served in the bureau of aeronautics and at Pensacola. He was a stark contrast to Browning, strong and fatherly instead of angry and demeaning. This was fortunate timing for the ship as they were about to embark on their biggest mission yet.

Celebrating Communion in forecastle on USS *Hornet* (CV-12). Chaplain Allen A. Zaun, May 28, 1944.

At the same time, the Japanese military was making plans of its own. They knew the Allies next move would be to invade somewhere that they considered to be a critical part of their inner defense ring, potentially in New Guinea, the Palau Islands, or the Caroline Islands. The original Operation Z (from August 1943) had fallen into Allied hands after Admiral Koga's plane crashed during a storm in the Philippines while moving his headquarters from the Palaus, killing him. The plans were found and transferred to the Allies via submarine and decoded by Nisei translators. The Allies thus had a general idea of the Japanese strategy to seek a decisive battle. Admiral Toyoda, Koga's successor, updated and modified the plan since they knew the Allies now had a copy of Plan Z. The Japanese new plan to deal with the next invasion was called A-Go. For Operation A-Go, The Japanese had combined their rebuilt carrier force with the bulk of the remaining surface fleet near their fuel sources in the East Indies. If the battle took place close enough to these bases, they would initiate Operation A-Go because the US fleet would be in range despite Japanese fuel issues. Admiral Soemu Toyoda placed Vice Admiral Jisaburo Ozawa in charge of this fleet. When the US invasion started, the Japanese would launch their fleet, combine it with land-based air in the target area and from other nearby bases and finally win the decisive victory they had been hoping for since the war started.

On paper, this plan looked like a reasonable one. The Japanese carrier group was of comparable size to Task Force 58, especially with land based air to help. With the addition of land-based aircraft, they should outnumber the Allied aircraft and these additional aircraft would be flying from "unsinkable aircraft carriers" on land (much as the Allies had done at Guadalcanal). The problem was that the difference in quality of both fighter aircraft and pilots was significant. The *Hornet* alone carried forty Hellcats, thirty-three Helldivers, and twenty Avengers at this point. New carriers were coming in regularly, increasing this already formidable number. These were experienced pilots with superior aircraft. The US Navy had two and a half years of lessons learned in conducting air operations, perfecting the use of radar and multi-squadron actions. After avoiding major operations for a while, the Japanese were putting everything in place for a major clash. If they ended up using Operation A-Go, it would be the first aircraft carrier battle in well over a year.

When the Allied forces under MacArthur began their attack on Biak, as part of the New Guinea campaign, the Japanese thought this was their opportunity. The Japanese considered Biak to be a critical part of their inner defense ring. Task Force 58 was not involved, though, so they did not initiate Operation A-Go because the targets were not lucrative enough. Instead the Japanese initiated Operation Kon, a surge of land based air and troop reinforcements to the area to try to destroy the invading force that way. The Japanese sent extra land-based aircraft to join the struggle and organized a troop convoy guarded by surface ships. There was some hope that if the amphibious task force and its escorts could be damaged enough it would draw out Task Force 58, allowing the Japanese to then initiate Operation A-Go. The US forces struggled to take Biak (they had underestimated how many defenders there were and the Japanese were starting to use better defensive tactics), and so the Japanese had some hope that this plan could work. A fleet of cruisers and destroyers led by Vice Admiral Naomasa

Sakonju was dispatched to Biak with reinforcements. This fleet was detected and intercepted by a cruiser and destroyer force led by Admiral Crutchley and, after suffering some damage, the Japanese fleet retreated. This battle was very similar in structure to the Battle of Savo Island, the Allied disaster right after the invasion of Guadalcanal and the Allied commander was even the same. However, Sakonju was not as bold as Mikawa had been in 1942 and allowed himself to be driven off easily when he very likely had a chance to win in a night battle.

Another fleet was readied, this time with the super-battleships *Yamato* and *Musashi*, to reinforce Biak. While this operation was underway, the Japanese learned of the US invasion of Saipan. Realizing this was the larger attack that was supported by Task Force 58 and thus the real opportunity for a decisive battle, the Japanese called their forces back from New Guinea and instead initiated A-Go against the US fleet near the Marianas. This area was originally considered outside the range of the Japanese fleet but they had recently decided to allow the use of unrefined oil from Borneo. It was a very dirty fuel (and the engineers of the ships despised it), but it did allow the fleet to reach the Marianas. Everything was now set for one of the decisive battles of the Pacific War. Two large Japanese fleets (Admiral Ugaki's surface force and Admiral Ozawa's carrier force) left their respective bases to rendezvous west of Saipan and attempt to destroy the US fleet.

The Battle of the Philippine Sea

Given the strengths and weaknesses of each fleet and their aircraft, the Japanese had created a tactically sound plan. They would keep their carriers too far for the shorter-ranged US aircraft to attack and would operate in shuttle fashion: launch from the carriers, attack the US ships, and then land on airfields in the Marianas. They would then refuel, rearm, and attack the US ships again on the way back

USS *Hornet* (CV-12) anchored in Majuro Harbor. Great example of dazzle camouflage paint scheme, May 29, 1944.

to their carriers. In the meantime, the large numbers of land-based aircraft in the Marianas would also join the attacks, hopefully giving the Japanese the advantage in numbers. The Japanese high command reasoned that the US ships would not be able to stray too far from the amphibious forces attacking the islands and thus the Japanese would be able to maintain the range they wanted. If they could drive off the US carrier fleet then the amphibious force would have to retreat as well.

Admiral Spruance was in overall command of the invasion but Vice Admiral Mitscher was in charge of the carrier groups. Mitscher's career in World War II before leading the fast carrier task force had been mixed. He was the commander of the USS *Hornet* (CV-8) during the Battle of Midway and his ship did not perform well. After this, he was given command of all the land-based air assets under Admiral Halsey in the South Pacific. Among other successes in a difficult campaign, the mission to shoot down Admiral Yamamoto's plane was carried out under his command. Thus his reputation had improved enough for him to be given command of the fast carrier force in January 1944. He immediately showed that he had learned a lot about carrier combat since 1942 with the devastating attacks on the Marshalls and Truk. Mitscher's plan was straight-forward. He would conduct preliminary strikes on the Mariana Islands to knock out their air power. This would be followed by air attacks to support the invasion of Saipan in conjunction with shore bombardment from the heavy surface ships. In the meantime, he would conduct searches to keep watch for a Japanese fleet sortie. His primary mission was to protect the transports and allow the invasion to proceed. Destroying Japanese ships was a second priority, although, of course, every pilot and captain would want to destroy any Japanese ships that ventured into the battle area. With his large fleet he was very confident that he could defeat any Japanese force sent his way. The stage was set for one of the biggest naval clashes in history and certainly the biggest aircraft carrier clash. With the exception of submarines, this would be another battle where the ships never saw each other.

What were the forces that were going to be facing each other in this critical battle? On the US side, there were multiple fleets with over 600 ships. The main force was Task Force 58 and this included a staggering number of ships and aircraft:

TF 58.1 (Rear Admiral Joseph J. Clark): USS *Hornet* (CV-12), USS *Yorktown* (CV-10), USS *Belleau Wood* (CVL-24), USS *Bataan* (CVL-29), three heavy cruisers, four light cruisers, fourteen destroyers.

TF 58.2 (Rear Admiral Alfred E. Montgomery): USS *Bunker Hill* (CV-17), USS *Wasp* (CV-18), USS *Monterey* (CVL-26), USS *Cabot* (CVL-28), three light cruisers, fourteen destroyers.

TF 58.3 (Rear Admiral John Reeves): USS *Lexington* (CV-16), USS *Enterprise* (CV-6), USS *San Jacinto* (CVL-30), USS *Princeton* (CVL-23), one heavy cruiser, four light cruisers, fourteen destroyers.

TF 58.4 (Rear Admiral William Harrill): USS *Essex* (CV-9), USS *Cowpens* (CVL-25), USS *Langley* (CVL-27), four light cruisers, fifteen destroyers.

These seven fleet carriers and eight light carriers carried around 900 aircraft. In addition there were seven battleships, four more heavy cruisers, and multiple destroyers that would participate, plus twenty-nine submarines in the area. There was also the huge transport fleet. The fact that this fleet was not as big as the one that had just landed troops at Normandy shows how impressive the advantage the Allies had in industrial strength.

On the Japanese side, they managed to pull together a formidable force:

"A" Force (Vice Admiral Jisaburo Ozawa): Fleet carriers *Taihō, Shōkaku, Zuikaku*, two heavy cruisers, one light cruiser, and nine destroyers.

"B" Force (Rear Admiral Takaji Joshima): Converted fleet carriers *Jun'yō,* Hiyō, Light carrier Ryūhō, one battleship, one heavy cruiser, and ten destroyers.

Van Force (Vice Admiral Takeo Kurita): Light carriers *Chitose, Chiyōda, Zuihō,* two super battleships, two battleships, four heavy cruisers, one light cruiser, and nine destroyers.

In addition, there were two oiler fleets with a total of six oilers and six destroyers. Finally, the Japanese had a total of twenty-four submarines. They had a total of 450 aircraft on their carriers plus another 300 or so in the Marianas

USS *Hornet* (CV-12) anchored in Majuro Harbor. Note the quad 40-mm anti-aircraft guns on the bow under the flight deck.

or at bases near enough to be transferred there during the battle. This meant the Japanese had five fleet carriers and four light carriers *v.* the US Navy's seven and eight respectively. They had around 750 aircraft *v.* 900. These were tough odds to be sure, but not insurmountable with a good plan and some luck.

On June 6, 1944, the *Hornet* and the rest of Task Force 58 left Majuro for the Marianas. This was overshadowed somewhat by D-Day on the shores of Normandy in France happening the same day. On June 9 and 10, out of range of Japanese search planes, the fleet refueled. On June 11, the first carrier strikes began hitting the islands. Devastating fighter sweeps were followed by full strikes attacking targets identified in months of pre-invasion photo-reconnaissance.

The battle plan started to go wrong for the Japanese immediately. Admiral Kakuta was in charge of the aircraft based in the Marianas and he had over 170 aircraft at the start of the battle. As Task Force 58 approached on June 11, Mitscher decided not to send in a dawn strike. Once this did not happen, the

RADM Joseph J. Clark (center), Capt. Sample (right), Cdr. C. H. Duarfield (left) inspecting crew of USS *Hornet* (CV-12), May 31, 1944.

Japanese assumed the first attack would come in the next morning and Admiral Kakuta's aircraft relaxed their guard. Instead, the Allies launched an afternoon fighter sweep against all the islands in the area and it was devastating. Pilots from the *Hornet* were assigned to the airfields on Guam and Rota. In a fierce air-to-air battle, the *Hornet* pilots shot down twenty-three Zekes (out of twenty-six Japanese planes in total they shot down for the day). Most of the Japanese aircraft currently in the Marianas were either destroyed in the air or on the ground. Losses recorded ran as high as 150 of the original 170. This was followed by strafing and bombing runs, destroying facilities in and around the airfields plus most of the surface ships in the area. Airfields are difficult to put completely out of action by air attack alone and this would be shown as the Japanese were able to bring in aerial reinforcements multiple times during the battle.

On June 12, 1944, Task Force 58 did launch a more typical dawn strike. Although the Japanese were expecting it, the crippling losses from the day before meant there was not much they could do about it. The US planes were unhindered in the air and their biggest enemy became anti-aircraft fire. The Avengers, Helldivers, Dauntlesses, and Hellcats bombed and strafed at will. At the same time, a convoy of transports and escorts was spied moving away from the Marianas. Several attacks were vectored onto this group of ships throughout the day and it was thoroughly wrecked. Over half of the original thirty ships were sunk and most of the rest damaged. As an example of the pace of operations in the early part of this battle, here are the *Hornet* air strikes on the 12th:

0520: Fighter sweep (Guam): Sixteen Hellcats. Strike 1A (Agana Airfield): Eight Hellcats, thirteen Helldivers and nine Avengers.
0710: Strike 1B (Agana Airfield): Ten Hellcats, fifteen Helldivers and eight Avengers.
0940: Strike 1C (Guam): Fourteen Hellcats, eleven Helldivers and eight Avengers.
1141: Strike 1D (Rota Airfield): Fifteen Hellcats, seven Helldivers and eight Avengers.
1344: Strike 1E (Agana Airfield): Nine Hellcats, nine Helldivers and eight Avengers.
1513: Strike 1F (Agana Airfield): Sixteen Hellcats, seven Helldivers and four Avengers.

On June 13, the pattern was repeated. Task Force 58 aircraft roamed at will, bombing and strafing everything they could find and shooting down any planes that dared to challenge them. Anti-aircraft fire was intense and was the primary causes of losses in the early part of this battle. In the end over seventy US aircraft would be lost to anti-aircraft fire over the course of the invasion. As usual, the Allies employed a vigorous air and sea search and rescue program and many of the downed pilots and aircrew were rescued.

RADM Joseph J. Clark during inspection of USS *Hornet* CV-12 crew in Majuro Harbor, May 31, 1944.

On June 15, Task Force 58.1, including *Hornet*, and Task Force 58.4, having been diverted northwest, attacked Iwo and Chichi Jima as it was obvious that reinforcement aircraft were being flown in through there for the Marianas. The task forces were under the overall command of Rear Admiral "Jocko" Clark. By this time, Mitsher knew the Japanese fleet was on the way (see below), but he still confidently sent this force off, knowing that they had time to conduct their mission and return before the Japanese fleet would be in range to offer battle. These strikes went on for two days and the air battles were intense. On the first day, ten Zekes were destroyed in the air, seven on the ground, twenty-one seaplanes were destroyed, and several freighters were sunk. The *Hornet* flew off an eighteen-plane fighter sweep against both islands, followed by a forty-one-plane strike against Chichi Jima and a small search flight. The only loss for the *Hornet* was one Helldiver and crew.

A *Hornet* pilot, Lieutenant Lloyd Barnard, became an ace in a day over Iwo Jima on his first combat mission. Here is the story of his air-to-air victories in his own words from the Aircraft Action Report: No. 1:

> I would estimate there were 30–40 Zekes in the air when we arrived over the target. We were at 15000 when I saw several Zekes making runs on some F6s below us at 10000. We pushed over after them, and, as we did so, we saw 8–10 coming in below us. I made a head-on run on one from above. I turned as I passed to see him blow up. Wings and debris went everywhere.

No. 2: 'I pulled up and missed one and a Zeke pulled in front of me at 9000 feet. I fired on him from 6 o'clock at the same level. He blew up and I went right through his fire'.

No. 3: 'After that one, I turned around and there was a Zeke on an F6s tail. I fired a full deflection shot from 9 o'clock, below, and he blew up. By this time they were blowing up all over the place'.

No. 4: 'From there I pulled around until I saw one on the water at about 200 feet altitude. I got it, level at 8 o'clock, and it pulled over into the water'.

No. 5:

> I climbed back up for altitude to 5000 feet and saw a Zeke above me. It was at 8000 feet and making an overhead run on an F6 at 6000. I followed it down to the water. It went into its run and pulled through faster than I did, so I went into a wing-over. Two more F6s closed and it turned inside them. Before they could bring their guns to bear on it, I pulled up in a high wing over and shot it down from 8 o'clock, above, 100 feet off the water.

On the second day, in bad weather, the Japanese aircraft were caught on the ground and at least thirty were destroyed. The *Hornet* flew off a fifteen-plane fighter sweep followed by a nineteen-plane strike, a twenty-six-plane strike (all against Iwo Jima), and a four-plane search. No planes or aircrew were lost by the

Hornet. An interesting event that also occurred was the transfer of 112 prisoners from one of the freighters that had been sunk the day before. They had been picked up by USS *Charrette* (DD-581) and were brought over to the *Hornet*. Against this, over the two days, both task forces lost only two planes shot down in air to air combat and two more by anti-aircraft fire. After this, they were called back down south to be in position when the Japanese task force arrived. Rear Admiral Clark seriously considered disobeying his orders and instead moving behind the Japanese fleet that was approaching the Marianas. He felt that the two task forces could trap the Japanese fleet and prevent any from escaping. In the end, especially because Rear Admiral Harrill of Task Force 58.4 would not go along with his plan, he followed his orders and rejoined the rest of Task Force 58. A curious what if considering the way the battle ultimately ended.

While all this was going on, the Japanese had been busy. Going back a few days, at first they were not sure if this was merely a carrier raid or a full invasion so their initial reaction was to start shuttling in land-based aircraft from other bases to reinforce the Marianas. However, once the shore bombardment started on June 13, 1944 (Mitscher detached some of his fast battleships, cruisers, and destroyers to supplement the shore bombardment force of older battleships), the Japanese realized this was a full invasion and Task Force 58 would have to stay on station for weeks or even months. Now they could finally make the decision to initiate Operation A-Go. Vice Admiral Ozawa left with his carrier fleet to rendezvous with Admiral Kurita and his surface force to engage Task Force 58. The two fleets rendezvoused on June 16 and refueled on the 17th. The Japanese were very confident of their plan and felt they had a great opportunity to inflict severe damage on the US Navy and turn the tide of the war.

Unfortunately for the Japanese, the glaring difference between the two submarine arms of the opposing navies was already impacting the battle. The Japanese had already been spotted by the US submarine picket force and their movements were being accurately reported. On June 13, 1944, *USS Redfin* observed Ozawa leaving Tawi Tawi. On June 15, *USS Flying Fish* had spotted part of the fleet leaving San Bernardino Strait. On the same day, *USS Seahorse* spotted another part of the fleet near Mindanao. During this time several Japanese oilers and destroyers were sunk. Meanwhile, the Japanese submarines, based in Saipan, were deployed to gather information and attack the US fleet. They were completely unsuccessful. By the end of the battle they had gathered no useful intelligence, had not scored a single hit on any US ships, and had lost seventeen of their own number. The US submarines were just getting started.

Now the stage was set. Task Force 58 was waiting to the west of the Marianas but they could not venture too far away due to their primary responsibility of protecting the transport fleet. The Japanese fleet approached and was able to dictate the range—in this case outside the range of US carrier-based aircraft. The wind was also in their favor, letting them steam as they liked and launch and recover aircraft while the US carriers had to turn into the wind each time. Admiral Ozawa was getting mixed reports from the Marianas—he thought he still had a sizable land based air force available and that they had inflicted some damage on the US fleet already. Neither of these things were true. Even as new aircraft kept shuttling in, they were destroyed almost as quickly as they arrived. This was the situation on June 18, 1944 as the two fleets approached each other.

A radio intercept that evening gave the US an approximate fix on the Japanese fleet and it was about 365 miles to the southwest. Mitscher wanted to try to force a night, surface engagement but Rear Admiral Lee, who commanded the fast battleships was not keen on the idea. He knew that the Japanese still held tactical advantages in night surface battles. Once this idea was turned down, Mitscher asked for permission to steam towards the Japanese to be in position for a dawn strike on the Japanese fleet but Admiral Spruance refused this request. This was also in line with the orders he had from Admiral Nimitz. He had read the original Japanese Plan Z and knew one of their tactics was to try to lure away the main force while a secondary force would go around the flank to strike the transports. He ordered Mitscher to stay in position close to the islands and the invasion force. Mitscher would have to yield the initiative to the Japanese. Everything both navies had learned about carrier warfare in 1942 said this was a bad idea. The side that got in the first strike nearly always won. However, Spruance was counting on his overwhelming numbers, superior fighter aircraft and radar to help him win a defensive battle (at least at first).

On June 19, 1944, both sides were ready. The US fleet was organized with a large screening force of Battleships and their escorts to the west of the carriers. The plan was to have the Japanese aircraft run this gauntlet of anti-aircraft fire and possibly "waste" their attacks on the battleships instead of the carriers. Meanwhile the carriers launched search aircraft and combat air patrols. This included fighter sweeps over the Japanese airfields again in the Marianas. The attack aircraft were readied to be launched to attack nearby airfields (and to be kept out of the way) and all the rest of the fighters were on call. The Japanese had launched their own early searches with the planes they had left in the Marianas and they spotted the US fleet quickly (by 5:50 a.m.). As soon as they did they launched all of their remaining strike aircraft from the islands to attack the US fleet. These aircraft were swarmed by Hellcats and wiped out. Thirty-five Japanese aircraft were lost in exchange for one Hellcat. This pattern would be repeated throughout the day. There would be no help for Ozawa from the land-based air units.

In the Japanese fleet, they had begun launching aircraft early. Because their aircraft carriers were organized in three different fleets, they launched strikes from these different groups and they reached the US fleet at different times during the day. The first strike had sixty-eight aircraft and it was picked up on radar very early. Once this radar contact was verified all the US carriers launched their dive-bombers and torpedo bombers as planned to get them out of the way and prepared the rest of the Hellcats. At 70 miles from the US fleet at around 10:30 a.m., the Hellcats swarmed this first contact and quickly shot down twenty-five of them for the loss of only one Hellcat. As the strike closed in on the US fleet, they were attacked by another group of fighters and sixteen more were shot down.

The remaining aircraft saw the battleship fleet and attacked it, hitting the USS *South Dakota* (BB-57) with one bomb. That was the only damage inflicted on an American ship from the first raid.

The second Japanese wave was detected at 11:07 a.m. and had 107 aircraft. These were intercepted sixty miles from the US fleet and at least seventy were shot down before they reached the US ships. The aircraft that got through actually attacked the carriers this time and scored some near misses with bombs and torpedoes but did not inflict any serious damage. By the time they had run the gauntlet of the fighters and anti-aircraft guns, ninety-seven of the 107 attacking aircraft had been shot down.

The third Japanese wave had forty-seven aircraft and came in from the north. They were intercepted forty miles out at 1 p.m., and while only seven were shot down, the rest did not press their attacks and inflicted no damage. Forty of these aircraft actually made it safely back to their carriers—the only sizable group of aircraft to do so.

The fourth wave of over seventy Japanese aircraft missed the US fleet all together and flew towards the islands to land and refuel in two groups. They did find some US ships on their way and attacked but inflicted no damage. Meanwhile, they were pounced on by Hellcats as they tried to land on the airfields and at least forty-seven were shot down and most of the rest damaged. It was during this fight that a pilot from the USS Lexington uttered the famous words "Hell, this is like an old-time turkey shoot!!"

To try to understand the pace of this battle on the 19th, coming after over a week of air operations by all the US carriers, we can examine the *Hornet*'s flight operations again:

0727: Flight #1 of eight Hellcats launched.
0953: Flight #2 of eight Hellcats launched for Guam.
1038: Strike 1A of twelve Hellcats, fourteen Helldivers, and seven Avengers launched for Guam.
1108: Strike 1B of three Helldivers launched for Guam.
1124: Flight #3 of four Hellcats launched.
1142: Seven more Hellcats launched to join flight #3.
1419: Strike 1C of eight Hellcats launched.
1436: Strike 1D of seven Hellcats, fourteen Helldivers, and nine Avengers launched for Guam.
1602: All aircraft recovered.

An Air Action Report from Hellcat pilot Ensign William H. Levering gives some idea of the events of the day, in this case in action against raid #4 trying to land on Guam (one Val destroyed in air, one Val probably destroyed in air):

> Ens. Levering saw a formation of Vals come in from the southwest, circle the field and start to land. He made a high-side pass at one plane from 7 o'clock. The plane

Hellcat about to be catapulted off USS *Hornet* (CV-12) in Kwajalian. Pilot Lt. Cdr. Morse, May 5, 1944.

> made a right turn, caught on fire and went in. He got on the tail of another Val at 500 feet, scissored with him, thought he hit him and that he was smoking, but is not certain of this. His attention was diverted at this time by three Zekes which he noticed coming down on him. An F6 came down and drove two of them off and Levering evaded the third.

The final tally for the day was around 350 Japanese aircraft of all types destroyed in exchange for thirty American planes (and many of these pilots were rescued). It was as devastating and one-sided as any major air-battle had been for some time. The only direct hit on the US ship was the USS *South Dakota* (BB-57) and she stayed in formation and kept firing her anti-aircraft guns all day. There were some casualties and minor damage from near misses.

While this absolute destruction of the Japanese air units was occurring, the US had been trying in vain to locate the Japanese fleet with search aircraft. The range was the primary issue. However, the US submarine force had made up for this in a major way during the day long battle. At 8:16 a.m., the USS *Albacore* surfaced near the fleet with a perfect firing angle on the new Japanese carrier *Taihō*. The fire-control computer failed but Lieutenant Commander Blanchard fired a spread

Pilot transfer from USS *McCall* (DD-400) to USS *Hornet* (CV-12). The transfer is occurring via Breeches buoy while the crew of the *McCall* looks on, June 15, 1944.

of six torpedoes by eye. Four missed and a fifth was heroically detonated by an alert Japanese pilot who dove his plane into its path. The sixth torpedo hit and this should not have been a major problem; however, it managed to rupture two aviation fuel tanks and fuel vapor began to spread through the ship. Meanwhile, the USS *Albacore* dived and escaped.

The USS *Cavalla* also managed to maneuver into a firing position on the Japanese fleet and her target was the *Shōkaku*. At noon, the submarine launched a spread of six torpedoes as well and this time three hits were scored. Fires broke out immediately and the carrier stopped. The ship was ordered abandoned and within minutes there was a huge explosion and the ship was blown apart. Over 1,200 men were killed. Just like the *Albacore*, the *Cavalla* escaped.

Meanwhile, for hours the *Taihō* had been operating normally while fuel vapors slowly spread through the ship. A damage control officer made the decision to turn on the ventilation system to try to clear the fumes but this instead made them spread throughout the ship even more. Finally, at about 2:30 p.m., almost six hours after being hit with a single torpedo, a spark ignited the fumes and there was a series of devastating explosions. This was quickly followed by a second series of explosions and Japan's mightiest new armored aircraft carrier sank, taking over 1,600 of her crew with her. So without US aircraft even finding the Japanese fleet, Ozawa had lost two of his three largest aircraft carriers and was down to about 150 of his original 450 aircraft.

During the night of the 19th and the morning of the 20th, both fleets steamed west. Mitscher hoped to close the range while the Japanese at first were just trying to maintain it. Even on June 20, Ozawa still wanted to fight. As he steamed away, US search aircraft sought him without luck for most of the day. Finally at 3:40 p.m., the Japanese fleet was spotted 275 miles away moving directly away from the US fleet. Mitscher decided this was his last and only chance to strike the Japanese fleet so he ordered an all out launch. As the strikes were getting ready, they learned the Japanese fleet was actually 60 miles further away than originally thought! This meant a strike at or beyond the range of US aircraft and a night landing or ditching on the return. Mitscher continued with his first strike but cancelled the second. Soon 226 US aircraft were flying towards the Japanese fleet. The *Hornet* launched a strike of fifteen Hellcats, fourteen Helldivers, and eight Avengers.

At right around sunset the US aircraft caught up with the Japanese fleet. Ozawa managed to put about seventy-five aircraft into the air to defend the fleet but only around thirty-five managed to intercept the US aircraft. They fought reasonably well indicating that the small number of planes were perhaps piloted by some of the better pilots remaining to the Japanese. The anti-aircraft fire was intense but the US aircraft pressed their attacks home. Two tankers and the carrier *Hiyō* were sunk, while the carriers *Zuikaku, Jun'yō*, and *Chiyōda* were all damaged as well as the battleship *Haruna*. Sixty-five Japanese aircraft were shot down in exchange for twenty US aircraft. *Hornet* pilots attacked the *Zuikaku* and scored several bomb hits and were originally credited with sinking her but she actually survived. Ozawa was now down to thirty-five planes but even at this point still wanted to fight and ordered his battleships and cruisers to seek a night surface action against the US fleet but this was called off after a few hours. He finally realized that he was not getting any help from land-based aircraft and that he had not inflicted much damage on the US fleet and so he sought permission to retreat and it was granted

Low on gas and now flying in darkness, the remaining US aircraft headed back for the fleet. As the got closer, Mitscher made the risky decision to turn on all the ship's lights on so the planes could find their way back. Spotlights shone into the sky and 5-inch guns fired star-shells. This was risky because it would make the ships easy targets for prowling submarines. There were no submarine attacks however and this decision probably saved many lives. With fuel running out and many aircraft damaged all aircraft were ordered to land on any available carrier. Many ditched, others crashed as they landed and others found themselves on unfamiliar carriers. When the final accounting was done, six Hellcats, ten Helldivers, and four Avengers were missing and most likely shot down in combat near the Japanese fleet. Seventeen Hellcats, thirty-five Helldivers, and twenty-eight Avengers were ditched or destroyed in crash landings. By the end of the following several days, many crewmen were rescued from the ocean meaning the final losses for this attack were "only" sixteen pilots and thirty-three crew men. This was amazing considering how many aircraft had gone into the water. The *Hornet*, by now considered to be a very lucky ship, lost twenty-one aircraft but only one pilot and two crew men.

The Japanese did not have a single ship with damaged engines so after this last attack they were able to steam west and escape without further damage. Although the US fleet chased them for a while they were also proceeding more slowly in order to search for and recover their pilots. The Battle of the Philippine Sea was over. Although the bulk of the Japanese fleet escaped, their carrier arm was crippled. Three carriers lost, three more damaged, and the air squadrons that they had patiently rebuilt over the past year wiped out. This would be the last significant carrier on carrier battle of the war. Although Ozawa would venture out one more time with his carriers it was not in a serious attempt to engage in a carrier battle but rather as a decoy force. The US commanders were somewhat disappointed but they should not have been. They had essentially destroyed the Japanese carrier arm and demonstrated that conventional tactics were useless against the fast carrier force at this point in the war. The invasion of the Marianas proceeded unimpeded by the Japanese Navy.

By the end of the Battle of the Marianas, the *Hornet*'s aircraft flew over 3,000 missions and destroyed 233 Japanese planes. This was a large portion of the record setting number of planes the *Hornet*'s pilots destroyed during the war. The *Hornet* had not been damaged by the enemy.

3

World War II
Part 2

Operations Leading Up to the Invasion of the Philippines

For the next several months, the fast carrier force and the *Hornet* were extremely busy raiding Japanese bases across the Pacific. While the rest of Task Force 58 returned to Entiwetok after the Battle of the Philippine Sea, Admiral Jocko Clark returned to Iwo Jima and Chichi Jima again. This was humorously referred to as Operation Jocko and many claimed that he had a personal vendetta against these islands since he attacked them so many times. On June 24, 1944, the fleet traded blows with the Japanese aircraft there, shooting down sixty-six aircraft. While several Japanese attacks were launched against the US ships there were no hits.

After a return to Entiwetok on June 27, 1944, Task Force 58.1 returned to Iwo Jima and Chichi Jima again on July 3 (fighter sweep) and July 4 (full attack), sinking a convoy of ships and destroying more aircraft. This was the most lucrative attack on these islands for the *Hornet*. Following this attack, the fleet moved back to the Marianas to continue to support the invasions there. For the *Hornet*, this meant attacks on Guam and Rota from July 6–21. This was followed by a quick return to Entiwetok and then attacks on Yap (July 25–26) and Ulithi (July 27–28). After another short trip back to Entiwetok, they went back to Iwo Jima and Chichi Jima again (August 4–5), and then returned to Entiwetok where they stayed for a while to rest, replenish and undergo a minor overhaul from August 23–28.

Aside from the heavy pace of carrier operations between late June and late August, many other significant events impacting the *Hornet* and her crew occurred. In August, Captain Sample was promoted and replaced by Captain Austin Kelvin Doyle. Captain Doyle was born in 1898 and graduated from Annapolis in 1920. He received his wings in 1922 and was then assigned to the USS *Wright* (AV-1). He was in VF-2 on the USS *Langley* (CV-1) and later was on the USS *Lexington* (CV-2), USS *Idaho* (BB-24), and was commander of the air group on USS *Saratoga* (CV-3) when World War II started. Before taking over the *Hornet*, he commanded the USS *Nassau* (CVE-16).

A cake for the 8,000th landing on the USS *Hornet* (CV-12) is given to Lt. (jg) Douglas E. Wynn by RADM Sample, August 6, 1944.

VT2 TBM Avenger with paining on the side indicating fifty-three bombing missions and three torpedo attacks for this aircraft, August 7, 1944.

USS *Cimarron* (AO-22) refueling USS *Hornet* (CV-12), August 2, 1944.

Rear Admiral John S. McCain took over for Clark on August 18, 1944 and Admiral Halsey took over for Admiral Spruance and the carrier fleet became Task Force 38 instead of 58 on August 27.

The *Hornet*, now part of Task Force 38, was back in action in September, launching strikes to support the upcoming invasion of the Philippines. This started an almost non-stop two months of fighting with attacks in the Palaus Islands (Peleliu and Malakal) on September 7 and 8, 1944, and in the Visayans (Islands in the central Philippines) from September 9–14. During this fighting, a Hellcat piloted by Ensign Thomas Cato Tillar was shot down over Leyte on September 12. After being rescued by Filipino guerillas, he was told there were not many defenders on Leyte or Cebu and once returned to the fleet he passed this information along. This intelligence resulted in the advancement of the timetable to invade the Philippines by two months by canceling several now un-needed supporting attacks.

By September, the *Hornet*'s VF-2 was the top-scoring fighter squadron in the navy and twenty-eight of the fifty fighter pilots were aces. However, the squadron was tired and they were replaced by Air Group Eleven on September 29, 1944. Air groups had been reconfigured to a new composition that included eighteen Avengers, twenty-four Helldivers, and fifty-four Hellcats. On October 4, the *Hornet* recorded its 10,000th carrier landing and this was followed quickly by riding out a large typhoon. Since leaving Pearl Harbor in March 1944, the *Hornet* had not returned home and repairs, replacements, upgrades, and the like were all handled at sea or at the advanced naval bases across the Pacific.

Battle board painted on the side of the island above some resting deck crewman on USS *Hornet* (CV-12) in Entiwetok, August 28, 1944.

VF-2 pilots with RADM Clark and Capt. Sample on the flight deck of USS *Hornet* (CV-12), July 1944.

Task Force 38 was next sent to Morotai in the East Indies (September 15–17, 1944), back to Luzon (September 21–24), Nansei Shoto/Okinawa (October 10–11), Formosa (October 12–15), and then back to Luzon again (October 18–19) in support of the immanent invasion of the Philippines. There is no intent to gloss over these raids, but they did not stand out in any serious way, except to the men involved. It was a deep dive into the inner Japanese perimeter and as we shall see, the Japanese tried to make the US pay for their boldness. Rather there is acknowledgement that they were similar in that outnumbered and overmatched Japanese land-based air units were overwhelmed by the large carrier attacks. Casualties did happen, though, from accidents, anti-aircraft fire, and occasionally enemy aircraft. The exception was when the carrier group struck Formosa.

The fighting around Formosa was fierce and deserves more detail. For the first time in a while there was significant damage to some US ships. Over several days, from October 12–15, 1944, a tough battle ensued. The Japanese had been able to keep aware of the movements of Task Force 38 and they knew about when the attack would come. This led them to activate their Sho-2 plan, another concentration of air power in the attempt to knock out Task Force 38. In Formosa alone, they had over 300 aircraft at more than a dozen airfields. At other airbases within range they could call in several hundred more. This battle ran day and night as both sides gave the battle everything they had. The Japanese even used a specially trained night torpedo squadron from Japan flying Betty bombers (called the "T" Force). At one point the fighting was so fierce that the US air force lent a hand with a large B-29 strike on Formosan airfields.

USS *Hornet* (CV-12) at sea after leaving Entiwetok Harbor, September 5, 1944.

Left: Ensign Richard B. Blaydes holding a cake signifying the 9,000th aircraft landing on the USS *Hornet* (CV-12), VF-2, September 12, 1944.

Below: Ten Hellcats with propellers turning preparing for a strike on Japan on an active flight deck, VF-11, October 10, 1944.

Over the course of the battle the *Hornet* air group shot down twenty-two aircraft (with another seven probable) for the loss of three Hellcats and their pilots. The *Hornet*'s anti-aircraft guns shot down two aircraft with a probable third and only suffered a near-miss bomb. Overall the Japanese lost over 500 planes, around forty cargo ships, and had all of their airfields heavily damaged. The US did not escape unscathed as they lost around eighty aircraft from all causes. More importantly, for the first time in a long time, US ships suffered serious damage as the USS *Canberra* (CA-70) was struck by a torpedo and the USS *Houston* (CL-81) was ultimately struck by two torpedoes. Both ships were towed out of the battle area successfully and in fact, Halsey tried to use them to lure Japanese surface ships out into battle. This almost worked as the Japanese had convinced themselves that they had done heavy damage to the US fleet (claims ran as high as eleven carriers sunk) and thought their surface ships could go out and pick off cripples. They announced the "destruction of Task Force 38" to their own people but eventually figured out that their pilots had grossly over-estimated the damage they thought they had caused and recalled their surface fleet.

Awkward Hellcat take-off during strikes on Formosa. Quad 40-mm Anti-aircraft gun in foreground, VF-11, October 12, 1944.

The Battle of Leyte Gulf

The Japanese knew the next invasion had to be either into Formosa, Okinawa, the Philippines, or the home islands. In any of those cases, if successful, the inner perimeter would be cracked and the Japanese would be cut off from their resources in the south. They laid out a new plan, called Operation Sho-Go, with multiple variants depending on where the next Allied invasion fell. The invasion of the Philippines set into motion a series of events that led to the next and last major battle between the Japanese and United States navies. Less than five months after the disastrous Battle of the Philippine Sea, the Japanese were ready for one last throw of the dice. They were ready to gamble the entire remaining fleet, just as they had in June, for a final chance to turn the tide. Even though the odds were stacked even heavier against them they actually came closer to succeeding this time. This battle would be different in that the Japanese had virtually no carrier air groups anymore so they had to devise a way to get their still significant surface fleet into fighting range of US transports to thwart the invasion.

The Japanese plan was very complicated as they usually were but given their desperation, they cannot be fully faulted for that. In this case, the complexity helped to cause confusion in the US response. They also thought they had at least another month before the next invasion so they were not fully ready when it started. Their plan was to bring ships together from all over the remaining empire into three distinct fleets: A northern decoy force composed of all of Ozawa's remaining carriers and their escorts (although they had very few planes) and then two large surface forces, one in the center, to come through San Bernardino Straight and one from the south to come through Surigao Straight. The carrier force was to lure Task Force 38 away from the transports supporting the invasion of Leyte so one or both of the surface fleets could get in among them and destroy them. Although the Japanese ships would have little naval air cover, they were promised land-based air cover. They knew this was unlikely to be enough but they were willing to accept heavy losses to succeed.

The US plan was simple: protect the invasion force and destroy any Japanese ships that showed themselves. Admiral Halsey was determined not to repeat the "mistakes" (in his opinion) that Spruance had made and would destroy the Japanese carriers if they showed up. Even if this meant temporarily leaving the transports since he felt that he could return to protect them before the Japanese would be able to cause significant damage. Even if the Japanese tried a flank attack like the one that Spruance had been so concerned about. While that flank attack had never materialized off of the Marianas as we will see, Halsey should have been more concerned about this possibility.

The *Hornet* ended up missing most of the Battle of Leyte Gulf. Task Force 38.1 had been dispatched back to Ulithi for rest and replenishment after weeks of intense operations. The fleet was well on its way to Ulithi when the Japanese fleet was first spotted by submarines steaming towards Leyte on October 23, 1944. Halsey ordered Task Force 38.1 to halt and refuel where they were instead

The *Hornet* maneuvering with other Task Force 38 ships in background during Formosa operations, October 12, 1944.

of going all the way back to Ulithi. That way they would be close enough to call back if needed. This enabled the *Hornet* to ultimately rejoin the battle after it was decided and help damage some of the retreating Japanese fleet.

Although this was the largest naval battle in history, it will be covered quickly here since the *Hornet* did not play a big role. US intelligence quickly surmised that the Japanese plan had multiple parts and a picture formed of the overall plan. US submarines quickly attacked Kurita's center force on October 23 and sank two heavy cruisers and damaged a third enough that it returned to base with two destroyers to escort. The next day, Task Force 38 air strikes began on Kurita's force and after a day of strikes, they sank the super battleship *Musashi*. Kurita turned his force around to the west, feigning retreat. By this time, the US had spotted Ozawa's decoy carrier fleet to the north and Halsey made the reckless decision to bring the entire Task Force 38 north with him, battleships and all, to destroy Ozawa's carrier force. This left San Bernardino straight undefended. Unknown to Halsey, Kurita had reversed course again, heading back towards the US landings. Thanks to Ozawa's sacrifice, this straight was undefended.

On October 25, 1944, the southern force came through the Surgiao straight and was met by a force of old US battleships (supported by cruisers, destroyers, and PT boats), including several raised from the mud of Pearl Harbor and was annihilated. Six of the first seven ships were sunk and the seventh damaged. The second group following behind turned around and retreated. Up north, Task Force

38 found Ozawa's carriers and proceeded to sink all four of them plus several of the supporting ships over the 25th and 26th. Here again Halsey almost made another mistake when he dispatched a small cruiser force to sink cripples and it almost encountered two Japanese battleships!

While all this was going on, Kurita steamed out of San Bernardino straight with the super battleship *Yamato* plus her escorting heavy cruisers, light cruisers, and destroyers against only a few escort carriers, destroyers, and destroyer escorts. What should have happened next was a complete annihilation of the US ships followed by destruction of transports. Instead, what followed was one of the epic naval battles of the war with the outnumbered and outgunned US fleet somehow inflicting severe damage on Kurita's ships and forcing him to retreat. Their urgent calls for help reached Halsey who belatedly tried to send them reinforcements.

Task Force 38.1 was asked to steam back toward the battle area to attack Kurita's force. On October 25, 1944, they steamed towards Leyte after finishing refueling, and late on the 25th, the *Hornet* finally entered the battle. Strikes were launched but the range was very long. The first strike had sixteen Hellcats, twelve Helldivers, and eight Avengers and ended up having a round trip of over 550 miles (one of the longest carrier strikes of the war for the US Navy) and a few Helldivers had to ditch. The fighters strafed some ships and many bomb hits were claimed on various ships but as far as can be told from Japanese records no bomb hits were scored, although there probably were several near misses. This was a good example of the optimistic claims often made by pilots. The *Hornet*'s after-action report indicated they scored two 1,000-pound bomb hits and one 250-pound bomb hit on a *Yamato*-class BB, four 500-pound bomb hits on a *Kongo*- or *Nagato*-class CA, two 500-pound bomb hits on a *Nachi*-class CA, one 1,000-pound bomb hit on a second *Nachi*-class CA, and one 250-pound bomb hit on a *Tone*-class CA from the first attack. At high speeds, it can be difficult to tell a near-miss from a hit. The second strike consisted of twelve Hellcats, nine Helldivers, and eight Avengers. The *Hornet* pilots claimed to have scored an additional two 500-pound bomb hits on an *Agano*-class CL and a 500-pound bomb hit on an unknown class CA from the second strike, although again there was no damage recorded.

On October 26, 1944, the *Hornet*, the rest of the task force, and other ships in range launched again and this time they obtained results. At first there was a search/fighter sweep with Hellcats. They found the Japanese fleet and scattered the few defending planes in the area. This was followed by three separate strikes and several bombs and torpedoes hit the *CL Noshiro,* sinking her, with the ship being credited to the *Hornet*'s flyers.

The final tally for the four separate battles that constituted the Battle of Leyte Gulf was devastating for the Japanese: one fleet carrier (*Zuikaku*—last of the Pearl Harbor attackers), three light carriers, three battleships, six heavy cruisers, four light cruisers, and nine destroyers were sunk. Against this the US lost one light carrier (to the first successful *kamikaze* attack), two escort carriers, two destroyers, two destroyer escorts, and a PT boat. However, the battle was very nearly a disaster for

Landing operations during Formosa strikes, October 12, 1944.

Ens. Charles H. Candon, fighter pilot in VF-11, accepts cake for making the 11,000th landing on the *Hornet*, October 1944.

the US Navy. Halsey made exactly the move he criticized Spruance for not making (going after the Japanese carriers with everything even at the risk of the transports) and it almost ended in a major defeat. In the end, the returning Task Force 38 would have most likely destroyed all of Kurita's ships if he had stayed but the damage he would have inflicted on the transport fleet would have been disastrous.

Operations in Late 1944 and Early 1945

Once this battle was over, Task Force 38.1 and the *Hornet* turned back around for Ulithi and their overdue rest and replenishment, which they reached on October 29, 1944. There was another minor overhaul done during this time in port. Otherwise, a normal stop with replenishment of all stores, replacement aircraft brought on board, and new crew joined.

Task Force 38.1 (*Hornet, Wasp, Monterey*, and *Cowpens*) was ordered to expedite their preparations so they could go back to sea. The Japanese air force, despite heavy losses, quickly became active again over the Philippines and Task Force 38.1 was ordered back out on November 2, 1944. Task Force 38.1

was tasked with covering central Luzon and *Hornet*'s Air Group Eleven was specifically to cover Clark Field.

This kicked off another intense period of operations for the *Hornet* and Air Group Eleven. The Japanese, despite their crushing naval defeat, were not ready to give up the Philippines. The critical impact of losing the Philippines had not changed of course. The Japanese kept trying to reinforce the island with more troops and were constantly flying in aircraft from all of their other bases in the area. Through most of November, the *Hornet* aircraft were heavily engaged above Clark Field and over Manila Bay, sinking many Japanese transports and small warships as well as destroying many aircraft. They were in combat on November 5, 6, 11, 13, 14, and 19. Finally after this grueling stretch of combat, they steamed back to Ulithi, arriving on November 22.

For almost three weeks, the *Hornet*'s crew rested, trained, and replenished. While they were doing this, the US naval commanders were giving deep thought to the new problem of the *kamikazes*. Multiple ships had been heavily damaged and sunk by the *kamikazes* and this new threat had to be countered. The first answer was to increase the number of fighters on board the carriers but there was a shortage of fighters and fighter pilots. It was decided to start training the Marines in their Corsairs for carrier operations. The first Marine squadron to join a carrier's complement was VMF-124 on the USS *Essex* (CV-9) in December 1944. The numbers of bombers was reduced to allow for the extra fighters. New strategies with fighter operations and radar pickets would be employed as well-resulting in more warning and larger contingents of fighters available for interception missions.

Christmas on the *Hornet* in 1944. Chaplain Zaun preparing for services while anchored in Ulithi, December 24, 1944.

On December 9, 1944, Keith Wheeler, of the *Chicago Daily Times*, and William Marien, from the *Sydney Morning Herald*, reported to the *USS Hornet* as war correspondents. Keith Wheeler was well known for his book "The Pacific is my Beat" which was focused on his time in the Aleutians earlier in the war. William Marien was to later write about the atomic bombing of Hiroshima for the Sydney Morning Herald. On December 10, the *Hornet* left Ulithi again to go back to the Philippines where the fighting still raged. It was during this trip that the *Hornet* passed 100,000 miles of travel. From December 14–17, the *Hornet* participated in multiple strikes over some of the same targets they had attacked in November. Using the 14th as an example, we can see how the configuration of the air strikes had changed. Note that "Picket CAP" consisted of fighters used to guard the picket destroyers and "Jacks" were special four quadrant, low altitude fighter missions specifically looking for low-altitude aircraft that might fly in under radar from the directions most likely for the Japanese to come from.

0630: Launched CAP 1 (Twelve Hellcats), Photo #1 (Four Hellcats), Strike 2A (Sixteen Hellcats), Strike 2B (Twelve Hellcats). All strikes had a target of airfields in the Bataan Peninsula and north of Subic & Manila Bays.
0852: Launched Picket Cap 2 (Four Hellcats), Strike 2C (Fourteen Hellcats).
1038: Launched Picket CAP 3 (Four Hellcats), Jacks 3 (Eight Hellcats), Strike 2D (Eight Hellcats) and Photo 2 (Four Hellcats).
1233: Launched Strike 2E (Eight Hellcats), Photo 3 (Four Hellcats) and CAP 4 (Twelve Hellcats).
1416: Launched Strike 2F (Six Hellcats).
1445: Launched Picket CAP 5 (Four Hellcats) and Jacks 5 (Eight Hellcats).
1657: Launched Dusk Jack (Four Hellcat night fighters).

Results of combat: two Vals and one Dinah shot down, all airfields strafed (destroying twelve aircraft on the ground), and one large transport damaged and beached.

On December 18, 1944, the fleet was caught in Typhoon Cobra (referred to by some as Halsey's Typhoon) and the *Hornet* survived this although many planes were damaged. The *Hornet* reported winds of over 70 mph from the storm that was centered about 40 miles away. Early in the morning the port lower boom was carried away by the winds. Overall, three destroyers and over 140 aircraft were lost to the Typhoon and many other ships were damaged. A total 790 US naval personnel were lost as well. The Japanese themselves had not managed to do this much damage to a US fleet since 1942. A board of inquiry was formed and while they judged that Admiral Halsey had an error in judgment by not ordering his fleet to scatter and seek shelter, they did not recommend disciplinary action. This caused the fleet to be sent back to Ulithi (arriving December 24) where the crew was able to spend Christmas. The *Hornet* stayed in port until December 30 when the whole task force steamed into the Pacific again.

Christmas on the *Hornet* in 1944. Christmas "tree" and Christmas mail at Ulithi, December 24, 1944.

Christmas music performed by *Hornet* crewmen in the hangar bay, December 1944.

In many ways, this action was much like the others: steam into Japanese-held waters and engage Japanese land-based aircraft and merchant/naval shipping and then return. What was different this time was that the fleet ventured all the way to the South China Sea. In addition to their "normal" targets on Formosa (January 3, 4, 9 and 10, 1945), and in the Philippines (January 6 and 7), they also struck French-Indochina (Saigon and Cam Ranh Bay on January 12), and Hong Kong and Kowloon (January 15). Large numbers of Japanese aircraft and ships were destroyed and photographs of future invasion sites such as Okinawa were secured. These strikes had more traditional groupings of fighters with bombers. This entire operation lasted until the fleet returned to Ulithi on January 26, 1945. During these operations, several carriers were hit by *kamikazes* although the *Hornet* retained her "lucky" status as she was not one of them. The *kamikaze* was proving to be a difficult tactic to counter and after almost a year of virtually no damage to US carriers, things were changing rapidly for the worse.

Back in Ulithi, Air Group Eleven was finally given a rest and transferred out on February 1, 1945. Air Group Seventeen joined the *Hornet* on February 2 and would remain for the rest of the war. Air Group Seventeen had been stationed on Guam and was happy to be aboard a ship now. After conducting some training exercises, they were deemed fit for carrier duty and the task force was ready to go back to sea on February 10. In the meantime, Manila had been liberated on February 3 and this essentially ended the navy's responsibility for the Philippines. Although no one knew it at the time, there were only two major invasions left for the navy to support: Iwo Jima and Okinawa. The fast carrier force's next several operations were all in support of these two invasions, both directly and indirectly. As usual, task force organization changed with almost every sortie and the *Hornet* was now in the company of the USS *Wasp* (CV-18), USS *Bennington* (CV-20), and *USS Belleau Wood* (CVL-24) under the command of Admiral "Jocko" Clark again, who flew his flag from the *Hornet*. Admiral Spruance had taken over the task force and thus it was designated Task Force 58 again. Each large carrier had at least seventy-three fighters in their new configuration and this meant they only had enough space for about thirty bombers.

The fleet left Ulithi on February 10, 1945 on their way to their next target: the Japanese home islands for Operation Jamboree. With the invasion of Iwo Jima to commence on February 19, the carriers were tasked with hitting any targets that could support Iwo Jima. In this case this meant a direct carrier strike on airfields and aircraft factories around Tokyo, the first since the Doolittle Raid in April 1942. The entire fleet consisted of eleven carriers, six light carriers, eight battleships, six heavy cruisers, nine light cruisers, and seventy-seven destroyers. They made their way north, skirting major island groups in an effort to arrive near the Japanese homeland undetected. Submarines and Destroyers went out ahead of the task force in order to prevent any detection by patrol boats. Large reconnaissance flights went ahead of the fleet as well. The weather was very poor and this helped. These tactics worked and the fleet arrived into launch position undetected by the Japanese.

Above: Avenger landing on deck during operations near Formosa, January 5, 1945.

Right: VF11 Hellcats preparing for take-off during operations near Formosa, January 5, 1945.

Left: Lt. Charles R. Simpson in his Hellcat during strikes into South China Sea. Note the sixteen air-to-air victories and the "Sundowners" insignia painted on the aircraft. He was the top scorer for VF-11, January 1945.

Below: Pilots in the ready room during operations in the South China Sea, January 1945.

The *Hornet*'s quad 40-mm anti-aircraft guns in action during strikes on Japan, February 16, 1945.

On February 16, 1945, Task Force 58 began their strikes on Japan. They launched over 1,000 aircraft towards Tokyo and surrounding areas. Due to bad weather, many aircraft had to hit secondary targets. Japanese aircraft were in the air but most seemed reluctant to engage. This was mostly due to orders to try to conserve their numbers for future operations. Airfields and aircraft factories were bombed. Thirty-six US aircraft were lost that day but only three for the *Hornet* (water landings with crews rescued). The *Hornet*'s pilots were credited with shooting down seven aircraft in the air and destroying many more on the ground.

On February 17, the process was repeated, although bad weather caused some of the strikes to be canceled. By the time the US Navy steamed away, the Japanese had lost around 100 planes in the air and another 150 on the ground, many ships sunk (including a CVE) as well as suffering severe damage to several airfields and factories. The *Hornet*'s pilots shot down two more planes in the air and scored hits on the Japanese CVE. The US lost sixty planes in the air and twenty-eight to accidents and had no damage to their ships. This was a very successful mission that brought the war home to the Japanese population in a different way than high-altitude B-29 missions.

Steaming away, the task force was in range of Chichi Jima on February 18, 1945 and sent strikes there to neutralize the airbase there before the invasion of Iwo Jima the next day. There was not a lot of opposition but the airfield was

Above: Helldivers of VB-17 on USS *Hornet* (CV-12), January 1945.

Left: CVG-17 pilots rushing to their planes on USS *Hornet* (CV-12), February 1945.

strafed and bombed. The next day began six days of operations against Iwo Jima in direct support of the Marines invading there. The results were mixed in that the napalm the aircraft were using was not as effective as hoped (many did not detonate) and the Japanese were so well dug in that it was difficult to get to them. There were some exceptions and in any event, the planes kept the Japanese from moving around above ground very much and did manage to cause some damage. The Japanese did not interfere much in the air except with night attacks on the fleet and of course the occasional *kamikaze*. The USS *Saratoga* (CV-3) was heavily damaged and the USS *Bismarck Sea* (CVE-95) was sunk during this time by *kamikazes*.

The fleet sailed north again for another strike on the Japanese home islands. On February 25, 1945, these attacks started in even worse weather than during Operation Jamboree. The weather was so bad that the strikes were called off after about half a day. The *Hornet*'s pilots claimed four Japanese fighters shot down in the air and damage done to airfields, planes on the ground and various small craft. Weather was turning out to be the Japanese homeland's best defense. The US Navy turned away for a planned attack on Nogoya the next day but was unable to reach their launching location in time and cancelled this attack too. Continuing to head south, the fleet reached a launching spot on March 1, 1945 to be able to attack Okinawa—the next invasion target for US forces. Important photo reconnaissance work was conducted as well as strikes against shipping and airfields. No Japanese aircraft contested this day's actions and all missions

VF-17 pilots rushing to their planes on USS *Hornet* (CV-12), February 1945.

VB-17 SB2C Helldivers flying towards USS *Hornet* (CV-12), February 1945.

were deemed successful. Many ships were sunk or damaged and all airfields were hit and damaged to some extent. After this the fleet retired to Ulithi for some much-needed rest and preparation for the invasion of Okinawa, scheduled for April 1, 1945.

The Battle of Okinawa

The invasion of Okinawa was going to be the navy's last big battle and would be the ultimate test of the fast carrier task force against massed *kamikaze* attacks. The battle at sea and on the island of Okinawa would let the Allies know that the invasion of the Japanese home islands was going to be extremely costly in terms of men and material. The rate of ship losses would have been a real challenge in an extended operation against Kyushu or Honshu. On the other side of the equation, Okinawa would also demonstrate that the Allied navies (the British contributed a large fleet for Okinawa) had such overwhelming numbers that victory was all but guaranteed. The only unknown was how high the price would be.

The fleet replenished and prepared for the invasion of Okinawa for ten days. The crew tried to get in as much rest and relaxation as they could during that time and they needed it. Not only had the pace of operations been high before getting back to port but the next mission was going to keep the *Hornet* and the other ships of the task force away from Ulithi (or any anchorage for that matter) for forty days. During that time the fleet was going to be subjected to the most massive waves of *kamikazes* yet in the final, true test of the new strategies and tactics of both sides.

Fully restocked and ready, the fleet left on March 14, 1945 to head back towards Japan to support Operation Iceberg, with the Marine and army troops to start hitting the beaches on April 1. Before this could happen, there would need to be suppression of Japanese air and naval units in the area and at bases within reach. In this mission, the fleet first steamed to Japan and started strikes on airfields in Kyushu starting March 18. Enemy air activity was heavy and in

VF-17 Hellcat wave off during landing operations on USS *Hornet* (CV-12), February 1945.

addition to defending their airfields they came out to attack the fleet. The *Hornet*'s pilots claimed thirty-four aircraft shot down for the loss of two planes (but only one pilot). Two of the *Hornet*'s pilots became aces in a day during that action. The *Hornet*'s records list twenty Zekes, three other fighter types, six bombers, three reconnaissance, and one transport plane shot down by the *Hornet*'s air-group for the day. Several more were destroyed by anti-aircraft fire. Several Japanese aircraft came close to damaging the *Hornet* but her string of luck continued and she was not damaged directly.

On March 19, 1945, the fleet had moved further north in order to attack the Japanese naval base at Kure. The USS *Wasp* (CV-18) was hit by a bomb and the USS *Franklin* (CV-13) was hit by two bombs that caught fueled and armed planes in the blast and the resulting fires and explosions crippled the ship. However, the *Franklin* was able to steam home and although she never fought again she was not sunk. In return, large numbers of Japanese aircraft were shot down and many ships damaged in the harbor. The *Hornet*'s pilots shot down another twenty enemy aircraft but lost six of their own. They reported the Japanese pilots were better trained than most they were used to meeting.

CVG-17 on USS *Hornet* (CV-12), May 29, 1945.

The fleet moved away from Japan's home islands, and on March 21, 1945, Japanese *kamikazes* came after them. A flight of sixteen Betty bombers carrying "Bakas" (as the Americans called them) were intercepted by the *Hornet*'s pilots and thirteen Bettys and one fighter escort were shot down. The Baka (called "Oka" by the Japanese) was a new rocket powered, piloted suicide aircraft that the Japanese were starting to use. Since it had such a short range it needed to be brought close to its target by a bomber. After it was released the pilot ignited the rocket engine and flew towards its target at over 500 mph. They were difficult to shoot down once launched but the bombers carrying them were vulnerable.

Starting on March 23 and lasting until April 5, 1945, the carriers struck targets across and near Okinawa, both before the Marines landed and after. Their targets were a mixture of airfields, shipping and targets directly supporting the ground troops, such as artillery emplacements, and defensive fortifications. In between here and there they might have a day "off" to refuel or to only patrol. However, this was all about to change as the Japanese were finally able to initiate their all-out *kamikaze* attacks. This went on from April 6 until late June 1945, although the intensity rose and fell. This was by far the most intensive fighting the navy has ever had to endure.

Over the course of April to June, the Japanese initiated their plan "Ten-Go" and were to launch ten massed *kamikaze* attacks against Allied ships. The Japanese called these attacks "*Kikusui*" which translates into "floating chrysanthemums." These ten attacks included more than 1,400 *kamikazes*. If you include the small, isolated attacks in between the large ones, the number of *kamikazes* used rises to over 1,900. During this time the Japanese launched some 5,000 sorties of *kamikazes* and standard attacks against the Allies. In April, there were four large attacks, followed by four more in May and two smaller ones in June.

Against this, the Allied navies had the largest concentration of naval air power ever seen. The fast carrier Task Force 58 had seven large carriers and six light carriers. The British had joined the fray with Task Force 57 and they added four large and six escort carriers. Protecting the transport and bombardment fleets were several task forces of smaller carriers, adding another fourteen to eighteen escort carriers. These were soon joined by marine squadrons operating from Okinawa itself as the first airfields were captured and put to use.

On April 6, 1945, the Japanese sent the largest *kamikaze* attack, no less than 355 *kamikazes* to attack the US fleet along with a large number of non-*kamikazes*. The British fleet was involved and they were primarily stationed to the south, near the Sakishima Islands, helping to suppress *kamikazes* by attacking the airfields there and on Formosa. The *Hornet*'s pilots shot sown at least forty-five of the attacking planes that day. Thus US lost three destroyers and three transports sunk plus many other ships damaged that day. This would set the tone for the big attacks to come. One *kamikaze* pilot bailed out and was captured and brought to the *Hornet*. He told is captors that there was another big attack planned for the 12th.

On April 7, the *kamikazes* continued their attacks and this time they managed to hit the USS *Hancock* (CV-19), damaging her. At the same time the Japanese sent the *BB Yamato* on a suicide run towards Okinawa (the so called First Diversion Attack Force) with an escort of a light cruiser and destroyers. Most of the ships of Task Force 58 joined in on the attack on this fleet. The *Hornet*'s pilots hit the Yamato with at least four torpedoes and two bombs and ultimately, after suffering over a dozen torpedo hits and at least twenty bomb hits, rolled over and then blew up and sank along with most of her escorts. With little to no air cover there was really no chance that this attack would succeed and it only resulted in the deaths of thousands of Japanese sailors.

The pattern of *kamikaze* attacks ebbed and flowed. April 12–13, 15–16, and 27–28 were all days of big attacks. Other days had smaller attacks. Spruance tried to keep at least two elements of Task Force 58 on station at all times while the third was pulled away to restock. The *Hornet*'s pilots shot down thirty-two aircraft on the 12th, eighteen on the 14th, nine on the 15th and as many as forty five on the 16th. On other days they were either refueling, patrolling only (the Japanese did not attack every day), or conducting strikes on Okinawa and other nearby islands. Finally the *kamikaze* attacks slowed down and eventually subsided, but before they did, they sank thirty-eight US ships and damaged 360 more. Almost 5,000 US sailors lost their lives and almost the same number were wounded. The *Hornet* was not directly hit. Some of the most severe fighting occurred around the US picket destroyers that were stationed in small groups far in advance of the fleet to give early warning of Japanese aircraft. Support ships were also hard hit.

After the *kamikaze* attacks ended, the *Hornet* and the rest of Task Force 58 stayed near Okinawa supporting the invasion until they finally steamed away on April 27, 1945. During that time, the *Hornet*'s pilots flew over 4,000 sorties, shot down 265 enemy planes, destroyed another 109 on the ground, and sank twenty ships. Finally, after forty full days at sea, most of those spent fighting, the fleet returned to Ulithi for another much-needed rest. While the invasion was never in jeopardy of being called off due to damage by the *kamikazes*, the amount of damage inflicted was a sobering wake up call for what would be involved in an invasion of the home islands.

After arriving back at Ulithi on May 1, 1945, the *Hornet*'s crew rested before what was to be their last combat mission in World War II. On May 9, they steamed back out to the Okinawa battle again. On May 12 they arrived, and for almost a month they conducted ground support missions, airfield strikes, and other normal combat operations off of Okinawa and southern Japan. This included the almost complete destruction of the Kumamoto aircraft assembly plant. Although there were several more large waves of *kamikazes* during this time, the *Hornet* was not impacted much and again the ship suffered no direct hits or serious damage. At the beginning of June, the *Hornet* dispatched some bombers to Okinawa to operate from land bases in direct support of troops still fighting there.

Finally the luck of the *Hornet* ran out. Not from the Japanese but from Mother Nature herself. On June 4, 1945, a typhoon was reported and Task Force 38.1

Typhoon damage to flight deck as seen from the front, July 8, 1945.

Typhoon damage to flight deck as seen from port side, June 5, 1945.

Close up of the port-side catapult showing the typhoon damage, June 5, 1945.

was inexplicably ordered into it by Admiral Halsey. On June 5, buffeted by 70–110-knot winds and huge 50-foot waves, the *Hornet* was struck by a huge wave as it was rising back up from a trough. The flaw that the engineers had warned about so long ago while designing the *Essex*-class carriers came to pass and the flight deck collapsed from the forward edge to frame four. This was over 24 feet of the flight deck. On June 6, they tried launching planes off the damaged bow but it was deemed too dangerous after a plane crashed. For the next six days, the *Hornet* launched aircraft off the stern, continuing to launch strikes and fly CAP missions. This was the first time this was done in combat by a carrier, although, as mentioned before, the *Essex*-class carriers were designed for this.

On June 10, 1945, the *Hornet* steamed away from Okinawa, and on June 13, it anchored in San Pedro Bay in the Philippines. Two days later, a much-rumored decision was officially broadcast to the crew: the *Hornet* would sail back to the USA for repairs. On June 19, in the company of a destroyer escort consisting of USS *Hickox* (DD-673), USS *Hunt* (DD-674), USS *Lewis Hancock* (DD-675), and USS *Marshall* (DD-676), the *Hornet* steamed out the Philippines, bound for Pearl Harbor, where she arrived on June 29. This was the first time the *Hornet* dropped anchor in Pearl Harbor since she had left on her first war cruise on March 15, 1944, fifteen months earlier.

On July 2, 1945, the *Hornet* left Pearl Harbor to travel to San Francisco without escort. On July 8, the crew was very happy to steam under the Golden Gate Bridge and back to Alameda Naval Air Station, docking at Pier #2. This ended sixteen months and nine days of uninterrupted service in the Pacific. The next day, the *Hornet* steamed over to Hunter's Point in San Francisco where she would undergo repairs and overhaul until September 11, 1945. Captain A. K. Doyle was relieved of command on August 1, 1945 and Captain Charles Randall Brown was named as the new commanding officer.

Captain Brown was born in 1899 and graduated from the Naval Academy in 1921. He served on several ships including the USS *Langley* (CV-1) before going to Pensacola for flight training. He received his wings in 1924. Before taking over command of the *Hornet* he was the captain of the USS *Kalinin Bay* (CVE-68). He fought in many of the major battles of World War II including the Marshalls, the Marianas, and Okinawa.

On August 12, 1945, a ship's open house was held on board and over 100,000 visitors were able to come aboard and see the veteran ship. In the meantime, Atomic bombs had been dropped on Hiroshima on August 6 and on Nagasaki on August 9. The Japanese signaled their unconditional surrender on August 15 and the surrender documents were officially signed on September 2 in Tokyo Bay. The USS *Hornet*'s World War II service was over.

As far the overhauls and refits that the *Hornet* received during World War II, it is difficult to find documentation on exactly what upgrades were made during her service in World War II. Since the *Hornet* never even returned to Pearl Harbor until damaged by the typhoon, she was never in a major shipyard for large modifications. What does seem likely was that the minor overhauls mentioned in the war diaries referred to small changes like additional 20-mm anti-aircraft mounts, upgrades to the radar and similar smaller modifications. For example, when launched, the *Hornet* had forty-four single 20-mm anti-aircraft mounts but by the end of the war she had fifty-five. Extra guns meant extra crew and this also meant minor, ongoing modifications to sleeping quarters to try to handle the overcrowding.

Even these minor modifications display the versatility of the *Essex*-class design. Once in combat, the main threats to the *Essex*-class carriers were determined to be enemy aircraft. So "field modifications" included increasing the number of fighters, anti-aircraft guns, and improving the radar where possible. These changes made the *Essex*-class carriers very difficult to damage through conventional means at the disposal of the Japanese in 1944. However, once the *kamikazes* started even these measures started to show the limitations of the *Essex* class. The best pilots, in the best fighter aircraft, guided by the best radar and supported by the best anti-aircraft guns could not stop every *kamikaze*. As was seen on other *Essex*-class carriers, oftentimes just one *kamikaze* getting through was enough to knock that ship out of action. If the war had continued to an invasion of the Japanese home islands it would have been a real challenge for the US fleet.

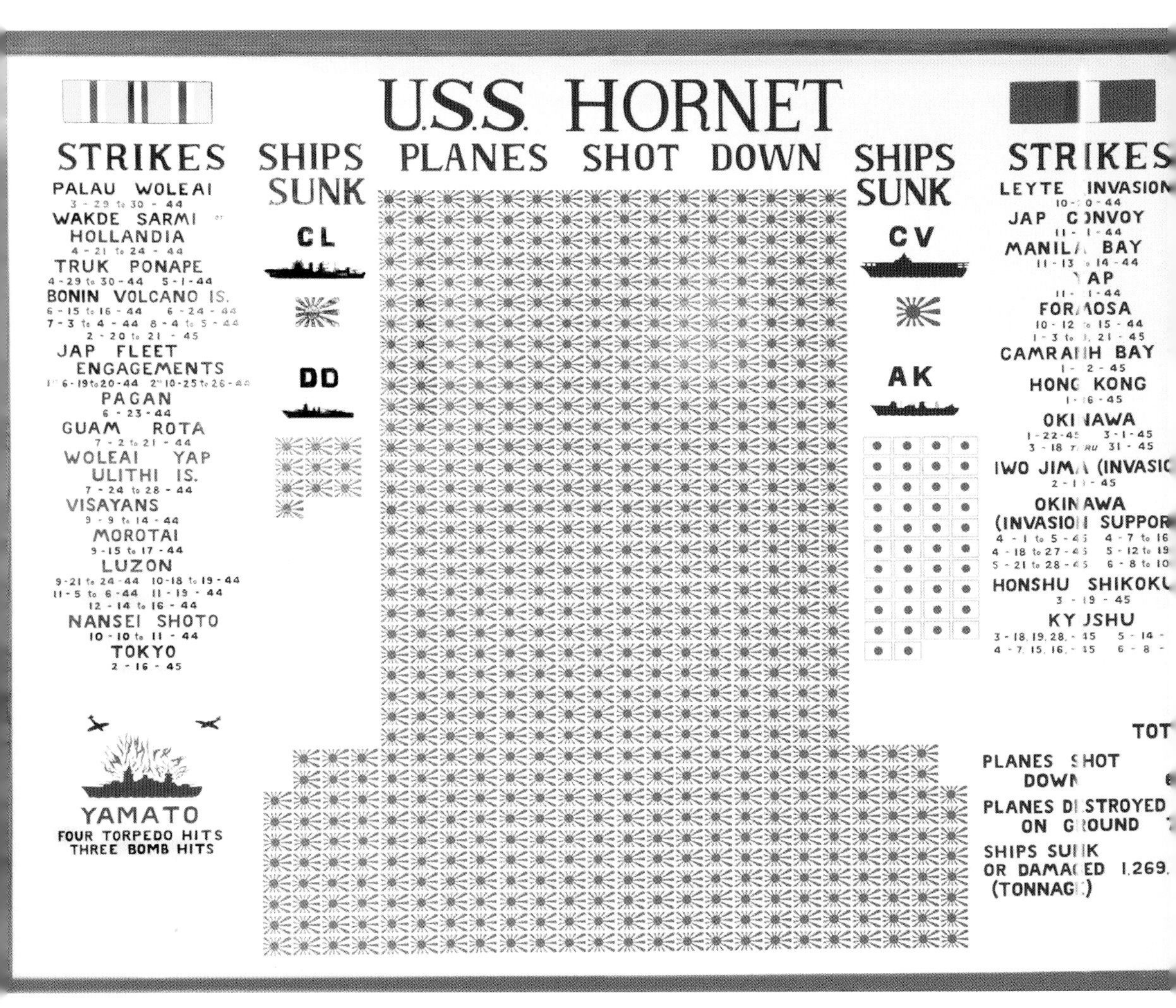

During World War II, the USS *Hornet* created quite a legacy:

Seven battle stars earned and the Presidential Unit Citation.
Enemy planes shot down: 668 (the most by any carrier in World War II).
Enemy planes destroyed on the ground: 742.
Enemy planes in one day: Sixty-seven.
Enemy planes in a thirty-day span: 255 (March 18, 1945 to April 16, 1945).
Number of pilots who became aces in a day: Ten.
Enemy ships sunk: Seventy-three (and another thirty-seven probable).
Enemy ships damaged: 413.
Nautical miles steamed: 155,000.

4

Post-World War II and Cold War

Operation Magic Carpet

With the end of the war, things would start to change rapidly. The *Hornet* was ready for action as all the upgrades were in place such as the new radar, extra 40-mm quad mounts, and a new portside catapult. The ship even had a fresh paint job. On September 12, 1945, the *Hornet* was underway in the eastern Pacific for post-repair trials and then back at Alameda Naval Air station on September 13. Some of the crew had been exchanged since there were sailors who had finished their tours when the *Hornet* returned to the west coast. This included Gene Millen who had served on the *Hornet* since she was commissioned.

The mission of the *Hornet* changed with the end of the war. Since the US had been preparing to invade the Japanese home islands, there were a large number of forward deployed soldiers, sailors, and other supporting personnel that had to be returned from the Pacific theatre. Operation Magic Carpet was initiated to bring them all home. The aircraft carriers were ideal for this because the hanger bay (with all of the planes removed) was enormous and could hold as many as 3,000 bunks. With added heads and washroom facilities the carriers could be quickly converted into large transports.

After the hangar bay had finished being modified, the *Hornet*'s first Magic Carpet mission started on September 18, 1945, when she steamed out of the Golden Gate bound for Pearl Harbor. Reaching Pearl Harbor on the 23rd, the *Hornet* loaded up and returned to San Francisco by the 30th. After a quick turn-around, the *Hornet* left on 2 October for the second Magic Carpet mission. Once reaching Pearl Harbor, however, the *Hornet* was retained for five days of carrier qualifications for pilots. Finally, on October 16, the *Hornet* left with over 800 Pearl Harbor dock workers bound for San Francisco.

On October 22, 1945, the *Hornet* embarked Air Group Nineteen for transportation to Pearl Harbor and left on the 26th. Instead of steaming directly to Hawaii, the *Hornet* stopped off in Monterey, California for Navy Day. October

Servicemen being processed in the hangar bay of USS *Hornet* (CV-12) in Pearl Harbor in preparation for returning to the United States during Operation Magic Carpet, October 16, 1945.

27 had been designated Navy Day because it was the birthday of noted supporter of the navy, Theodore Roosevelt. Some 30,000 people came aboard to see the ship before the ship finally left on the 29th. After disembarking Air Group Nineteen, the *Hornet* loaded passengers for the return trip. On the trip home, the *Hornet* was caught in a severe storm but eventually made it to San Francisco with minor damage. No doubt the passengers were happy to offload.

The fourth and fifth (and final as it turned out) Magic Carpet missions the *Hornet* made were all the way to Guam. Starting in November and finally ending in January 1946, the *Hornet* made two of these longer trips, once returning to Seattle and the final one back to San Francisco. Once back in San Francisco, the *Hornet* went on a short CARQUAL cruise and then returned to Alameda. Capt. Charles Frederick Coe relived Capt. Doyle as the commander in February 1946.

Captain Coe was born in 1901 and had graduated from the Naval Academy in 1923. He served on the USS *Florida* (BB-30) and USS *Litchfield* (DD-336) before enrolling in flight training. He earned his wings in 1926 and served on the USS *Maryland* (BB-46) and USS *Saratoga* (CV-3). After several staff positions and air group assignments, he was in command of patrol wing two at Pearl Harbor on December 7, 1941. He was highly decorated and served in multiple positions throughout World War II. He took command of the USS *Puget Sound* (CVE-113) in 1945 before taking over the *Hornet*.

Aircraft on the flight deck of USS *Hornet* (CV-12) in Pearl Harbor being brought back to the United States during Operation Magic Carpet. The aircraft are F4U Corsairs and F8F Bearcats, October 16, 1945.

Two aircraft carriers that survived all of World War II—USS *Saratoga* (CV-3) and USS *Enterprise* (CV-6)—with two built during the war—USS *Hornet* (CV-12) and USS *San Jacinto* (CVL-30)—at Alameda Naval Air Station in preparation for another Operation Magic Carpet mission, November 16, 1945.

USS *Hornet* (CV-12) returning from her final Operation Magic Carpet mission (from Guam) at Alameda Naval Air Station, January 28, 1946.

With the end of World War II and the end of Operation Magic Carpet, the navy had almost 100 aircraft carriers in service that it needed to decide what to do with. Many ships under construction were cancelled, and many others were slated for the reserve fleet. This list included the *Hornet*. The process of inactivation overhaul was begun in March 1946. Put into dry dock, the ship was inspected and everything that needed repair was taken care of. After this, everything was prepared for deactivation. Protective coating was applied to many surfaces, dehumidifiers were installed, and skeleton crew quarters were readied. On August 14, 1946, the *Hornet* was officially placed in "Inactive status: In Commission, In Reserve." At this point, Cdr. Karl E. Jung relieved Capt. Coe. Jung graduated from Annapolis in 1931. Once he finished his flight training, he was assigned to VF 3-F. He participated in an exercise on the USS *Ranger* (CV-4) and among other positions served as commander of Air Group Nineteen on the USS *Lexington* (CV-16) in 1944 before taking over the *Hornet*.

On January 15, 1947, the *Hornet* was officially placed in "Inactive Status, Out of Commission, In Reserve." The last of the crew was reassigned and the *Hornet* joined the USS *Intrepid* (CV-11), USS *Shangri-La* (CV-38), USS *Antietam* (CV-36), and USS *Independence* (CVL-22) in the San Francisco "Mothball" fleet. The *Hornet* would remain here for four years.

This time period showed once more the versatility of the *Essex* class. The ability to quickly turn the large hangar bay into a troop berthing area was very useful and meant the navy was saved from finding a larger numbers of extra transport ships. The best thing about this modification was that it was also easily removed—converting these ships right back into aircraft carriers just as fast.

Cold War: The Korean War and the SCB-27A Refit

The *Hornet* was not destined to stay in the reserve fleet for long. Since the end of World War II, the navy had been considering how best to maintain their aircraft carrier force. Should they build new carriers with new specifications? Should they update the large number of older carriers to save money? Or should they do a combination of both? Given budget constraints, they opted to try to do both. During the late 1940s as the *Midway*-class carriers were coming online and newer designs were being considered, there was a lot of discussion on how to use the large number of *Essex*-class carriers still available. A design proposal was put together and was first tested with the USS *Oriskany* (CV-34), an uncompleted *Essex*-class carrier that had its construction suspended in 1946. The construction was then restarted and completed in 1950 with the basic upgrade all *Essex*-class carriers would get based on the need to operate jets and incorporating all the lessons learned in World War II. This was the SCB-27 upgrade.

The situation in the Pacific was rarely peaceful in the 1950s. The first major Cold War conflict between the Communist and Democratic powers in the Pacific area was the Korean War. After the Communists won in China in 1949, they signed a thirty-year treaty with the Soviet Union in February 1950. Backed by both the Soviets and the People's Republic of China, the North Koreans were supplied, supported, and encouraged to reunite Korea as a Communist country. Once the Korean War broke out in 1950, the urgency to bring carriers back into service and the related budget dollars needed were both heightened substantially. Various *Essex*-class carriers were designated for the SCB-27 upgrade and were brought into shipyards to begin this work. Eventually seventeen of the nineteen mothballed *Essex*-class carriers were reactivated between 1950 and 1955. There were slight variations on SCB-27 and the *USS Hornet* was slated to receive the SCB-27A upgrade.

On March 20, 1951, the *Hornet* was recommissioned under the command of Captain Francis Lee Busey and moved from San Francisco to the New York Naval Shipyard. Busey had a long navy career that started when he graduated from the Naval Academy in 1926 and included operations in North Africa and the South Pacific in World War II.

The *Hornet* was then decommissioned on May 12, 1951 for conversion to CVA-12 (the A stands for attack and was introduced into the navy nomenclature in 1952), under Commander Gorman Coady Merrick. Merrick graduated from the Naval Academy in 1934 and was granted his wings in 1938. His career in the navy included service on the USS *Lexington* (CV-2). During World War II, he served as the commanding officer for the USS *Thrush* (AM-18) for one year.

While USS *Hornet* (CV-12) was undergoing the SCB-27A refit, USS *Wasp* (CV-18) was damaged in an accidental collision with a destroyer. The bow was removed from the *Hornet* to use on the *Wasp* at New York Shipyard, Brooklyn, May 8, 1952.

The actual construction work started in July of 1951. While this was happening, the USS *Wasp* (CV-18) collided with another ship and damaged its bow severely. To get it back into action quickly, the bow from the *Hornet* was cut off and welded onto the *Wasp*. A new bow was then constructed and put on the *Hornet*. Finally the *Hornet* was commissioned as CVA-12 on September 11, 1953. The refit had cost $50 million (compared to $69 million for the original cost of the entire ship). It took over two years to complete, longer than it had taken to build the entire ship the first time.

SCB-27A

What did the navy get for their $50 million investment? Quite a lot as it turns out. The *Hornet*, like her sister ships, was radically changed to function in the new era of jet aircraft:

1. In order to handle heavier jet aircraft, the flight deck was heavily reinforced. The elevators also had to be made bigger and stronger with the two centerline elevators expanded to 44 feet by 58 feet with a weight capacity of 40,000 lb. The third, side elevator was only expanded to handle 30,000 lb. The idea was

USS *Hornet* (CVA-12) *en route* to Guantanamo Bay with ATG 181 on board, January 10, 1954.

that aircraft could be fueled and armed in the hangar bay and then raised to the flight deck on the two large elevators while the third elevator could lower them when they were lighter after using fuel and expending their armament.

2. To facilitate jets taking off and landing, two new H8 slotted tube hydraulic catapults, retractable jet blast deflectors, and a new Mk. 5 arrestor gear set were all installed.
3. To facilitate the fuel and armament needs of newer aircraft, the aviation gas storage was increased to 300,000 gallons, an increase of 50 percent and the corresponding fuel pumps were increased in capacity. The bomb elevators were expanded and strengthened, with the forward bomb elevator in particular having the capability to handle a 15-foot-long, 16,000-pound package (allowing for lifting nuclear weapons).
4. Anti-aircraft armament was significantly altered. Gone were the 20-mm and 40-mm mounts. The deck 5-inch guns were also gone. Now there were eight 5-inch guns, four on each side just below the flight deck. In place of the smaller anti-aircraft guns were the fourteen of the newer 3-inch/50-caliber twin mounts, also on the sides just below the flight deck.
5. With all the guns removed from the island and the flight deck, the island itself was heavily redesigned. It was taller but shorter in length. It had one large mast for the radar and communications. The boiler uptakes were angled aft.
6. Damage control lessons learned from World War II led to several safety based redesigns: the hangar bays were separated into three compartments with large, blast/fireproof bulkheads that could be closed to seal off the hangar bays from each other. Firefighting capacities were heavily increased with a fog/foam firefighting system, better fire mains and improved water curtains. Weapon stowage and handling facilities were improved. The ready rooms for the pilots were relocated below the armored hangar bay for more safety and an escalator was added on the starboard side to quickly get the pilots to their aircraft.
7. Other improvements included increasing the electrical generating power from 250 kilowatt to 850 kilowatt, an improved radar system, and a larger, stronger crane added on the starboard side.
8. All of these improvements led to a substantial increase in weight. This was compensated for (somewhat) by removing the side armor belts. This in turn was compensated for by adding more blisters to the side. Still, weight increased by 20 percent and speed was decreased to 31 knots. The waterline beam ended up expanding by 8 feet to 10 feet.

Admiral Halsey (retired) addresses the assembled officers, crew, and guests at New York Shipyard, Brooklyn. The World War II battle board can be seen in the background, September 11, 1953.

During the recommissioning ceremony of USS *Hornet* (CVA-12), Capt. Milton A. Nation addresses the assembled officers, crew, and guests at New York Shipyard, Brooklyn. The World War II battle board can be seen in the background, September 11, 1953.

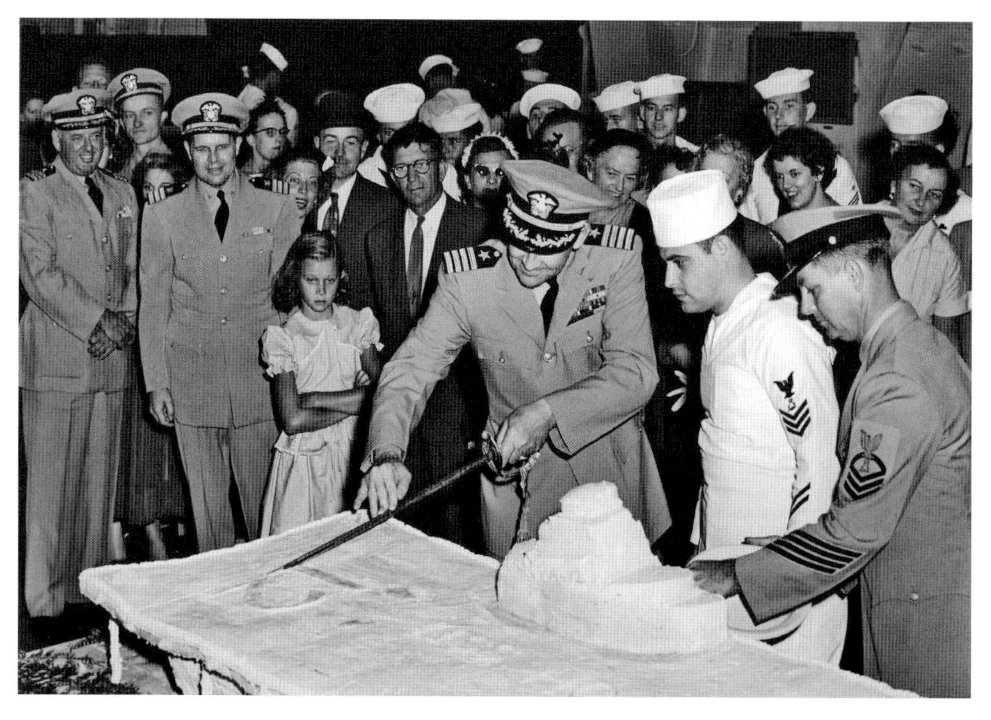

Capt. Milton A. Nation cuts a cake in the shape of the *Hornet* during the recommissioning of USS *Hornet* (CVA-12) at New York Shipyard, Brooklyn, September 11, 1953.

CVA-12 and Cold War Tensions

On November 8, 1953, the *Hornet* started two days of sea trials before heading down to Virginia. On November 18, the *Hornet* started two weeks of shakedown in the Virginia Cape area with Air Group 181. On December 8, the first jet (an F2H Banshee) landed on the *Hornet*, piloted by Lt. (jg) R. G. Wallace. The *Hornet*, now under the command of Capt. Milton Adolphus Nation, headed down to the Caribbean for a full workup in January 1954. Capt. Nation was born in 1904 and graduated from the Naval Academy in 1927. Of note, he received the Navy Cross for his bombing missions against the Vichy French fleet off Morocco from the USS *Suwannee* (CVE-27). Every day they steamed out of Guantanamo Bay to practice all aspects of air and ship operations. After returning to Norfolk, the *Hornet* disembarked Air Group 181 and embarked Air Group Nine. Aircraft included the F9F-5 Panther, F9F-6 Cougar, F2H Banshee, and A-1 Skyraider.

Although the urgency of the Korean War had accelerated the navy's plans to reactivate many mothballed ships, the Korean War actually ended in July 1953, before the *Hornet* was recommissioned. Given this, the navy decided to have the *Hornet* join the Pacific Fleet the "long way" since they were located on the East Coast. That meant heading east instead of west, via the Atlantic, Mediterranean, and Indian Oceans. This was the start of an eight-month cruise that crossed the Atlantic and first stopped in Lisbon, Portugal. After entering the Mediterranean,

USS *Hornet* (CVA-12) steaming down the East River in New York before her sea trials following the SCB-27A refit, October 31, 1953.

USS *Hornet* (CVA-12) steaming past the incredible skyline of New York City in preparation for her sea trials following the SCB-27A refit, October 31, 1953.

they stopped at Naples, Italy, and Port Said, Egypt, on their way to the Suez Canal. In the Indian Ocean, they stopped at Colombo, Ceylon (now Sri Lanka), and then to Singapore in Malaya. Finally the *Hornet* steamed into Manila in June 1954 and officially joined the US Seventh Pacific fleet. The last time the *Hornet* was in Manila was during World War II and several of the ship wrecks in the area had been sunk there by *Hornet* aircraft.

Lt. (jg) Dale Berven and A. N. Joseph Leather Jr. were two crewmembers who served on the *Hornet* during this around the world cruise. Both had served in the Korean War on the USS *Philippine Sea* (CV-47) before transferring to the *Hornet* for its first cruise as a jet-carrying carrier. Berven shared:

> I was in VF-91 and I flew the F9F-6 Cougar. I flew it from Alameda NAS all the way to Norfolk in several hops. Once I got to the *Hornet* I quickly learned two things. The first was that the food was great! The second was that the ship was not very welcoming to us. For the first time I discovered that there was a rivalry between the east coast and the west coast Navy, not to mention between the black shoes and brown shoes [black shoes were airmen and brown shoes were sailors].

Bombs being lifted from a lighter to USS *Hornet* (CVA-12) by the ship's crane as she prepares for active duty in Virginia, November 19, 1953.

Ammunition being loaded onto USS *Hornet* (CVA-12) as she prepares for active duty in Virginia, November 19, 1953.

The first jet landing on USS *Hornet* (CVA-12). The aircraft was an F2H-3 Banshee piloted by Lt. R. G. Wallace (VF-41). The two cannon ports on the left side of the aircraft can be clearly seen, December 8, 1953.

Commander in Chief of the Atlantic Fleet ADM Lynde D. McCormick salutes as he prepares to board an aircraft to leave USS *Hornet* (CVA-12) in Guantanamo Bay, Cuba, January 19, 1954.

As for the cruise itself, Dale remembered:

> Our first stop was Lisbon, Portugal and it was a very nice, clean city. The people there were very friendly. Our next stop was Naples Italy and I remember it being dirty in comparison. There were good tours though and I was able to visit Pompeii and Rome while there. Next was a trip through the Suez Canal. On the port side was empty desert while on the starboard were roads, houses, greenery and British soldiers. I remember one British soldier yelling to me, "When are you Yanks going to get a bigger ship?" This was pretty funny since I think at that time the *Hornet* was the biggest ship to have passed through the Suez Canal! It took us about fifteen hours to make it through the canal and while in that area it was so hot that many of the crew slept on deck. Because we weren't at war there was only limited flights happening-usually just a four plane CAP so all of our flight hours went down.

Joseph Leather Jr. was a plane captain in VF-94. VF-94 flew F9F-5 Panthers and he was responsible for one of them. As an enlisted man, his memories and experiences are a bit different from Berven's. Leather said:

During the 1954 world cruise, USS *Hornet* (CVA-12) refuels from USS *Severn* (AO-61) *en route* to Naples, Italy. A large number of CVG 9 aircraft on deck can be seen, May 27, 1954.

> I don't remember the ship not welcoming us but we did keep to ourselves. I flew from Alameda Naval Air Station on a transport. There was something that not a lot of people knew and that was that we carried nuclear weapons back then. It was kept pretty quiet and while we did load them onto aircraft from time to time we never launched an aircraft armed with a nuclear weapon. In Naples I remember the kids there would try to get your shoes dirty so they could sell you a shoeshine! I switched my day off the ship so the Catholics could go see the Pope. I did get off the ship to see Rome though.

Once in Colombo, Ceylon, their experiences were different again. Because Berven was an officer, he had more privileges. Berven said:

> In Colombo I went to the British Officer's Club there and met Major Henry Ashworth, British army officer. After talking for a while I invited him and his family on board for a tour of the *Hornet*. He had his wife Gwen and two daughters with him and they really enjoyed their tour and dinner on board.

Leather remembered, "I went on a tour and saw giant bats and elephants getting baths."

As they moved into the Pacific they both had similar recollections of Hong Kong, Manila and Yokuska. Berven said:

> In Hong Kong you could get very inexpensive suits. In Manila you could see lots of shipwrecks everywhere. In Yokuska, the exchange rate was very favorable, perhaps

USS *Hornet* (CVA-12) transits the Suez Canal, June 5, 1954.

> 360 to one for the Yen to the dollar so the sailors could afford all kinds of things, including enjoying the company of women there.

During this time and over the next several years, the People's Republic of China (PRC) instigated several international crises. In general, the PRC wanted to reclaim all Nationalist Chinese-held islands off the coast of China, especially Taiwan (known as Formosa in World War II) and more generally encourage the spread of Communist governments. On the US side, the major democracies of the Pacific signed the Manila Pact in September 1954 establishing the Southeast Asia Treaty Organization (SEATO) with the express purpose of limiting the expansion of Communism in the Pacific. The signatories included the US, Britain, Australia, New Zealand, Thailand, the Philippines, Pakistan, and France. This supplemented US defense pacts already in place with the Philippines and South Korea. As the role of the Seventh Fleet in the Pacific increased, the need to build up and expand naval bases in the area became more urgent. Major bases for the US Navy included Yokosuka and Sasebo in Japan (including Atsugi NAS), Okinawa (Kadena NAS), and Subic Bay in the Philippines with Cubi Point and Sangley Point NAS.

A F9F-6 Cougar is on the deck edge elevator with wings folded while at sea in the Pacific with Seventh Fleet approaching Manila, September 5, 1954.

On July 19, 1954, Capt. Frank Alvin Brandley took over the *Hornet*. Brandley was commissioned in 1929 and served on the USS *Colorado* (BB-45) before taking flight training that he completed in 1930. He served on the USS *Lexington* (CV-2), in Panama as part of a patrol squadron, on the USS *Tuscaloosa* (CA-37) leading up to World War II. In the war he served in various capacities in patrol squadrons in the Pacific, and as the executive officer of the USS *Boxer* (CV-21). After World War II he commanded an air group and then the seaplane tender USS *Suisun* (AVP-53), USS *Kula Gulf* (CVL-108), and finally the *Hornet*.

Almost immediately the *Hornet* was caught up in a politically tense situation. On July 23, 1954, two People's Liberation Army Air Force (PLAAF) fighters (Soviet-supplied Lavochkin or LA 11s) shot down a Pacific Cathay C-54 Skymaster passenger plane near Hainan Island. This incident is often referred to as the "Pacific Cathay Douglas DC-4 Shootdown" because the passenger plane version of C-54 was called the DC-4. It is still unclear why this happened (although the Chinese had been harassing reconnaissance flights and may have thought the DC-4 was a spy plane), but as the aircraft was going down the pilot sent out multiple distress calls and both the British and US navies responded by sending search aircraft to the area. In addition, the US Navy sent a small fleet that included the *Hornet* and the USS *Philippine Sea* (CVA-47) to help in the search. The nine surviving passengers were all rescued but search operations continued for a few more days in the hopes of finding more survivors or bodies. On July 26, two PLAAF LA-11s attacked some of the search aircraft and were shot down by *Philippine Sea* Skyraiders and a Corsair. On this same day the Chinese government admitted their responsibility for shooting down the passenger plane and the situation de-escalated.

Berven was flying during this time. He said:

> I didn't see any Chinese aircraft but I was up there flying CAP. I do know that after the incident, when the A-1s shot down the Chinese fighters, *LIFE* magazine was on board our ship and they used a picture of us when they wrote about the incident. The flyers on the Philippine Sea were so angry! Even worse, the picture they used of us was of all of the junior grade Lieutenants so the regular Lieutenants on the *Hornet* were also angry!

On May 7, 1954, Viet Minh forces defeated the French at Dien Bien Phu and by July a ceasefire was signed, ending the conflict. Part of the ceasefire agreement was to divide the country in two and allow all forces and civilians to move to whichever part of the country they wanted to be a part of. At roughly the same time, the PRC was escalating tensions in the Taiwan Straight by attacking Nationalist Chinese-controlled islands. A mutual defense treaty had been signed between the US and Nationalist China and the Seventh Fleet was sent to the area as a show of force. The *Hornet* was not in the Pacific at this time.

By January 1955, the *Hornet* was back on the West Coast and spent until May 1955 steaming up and down the west coast from San Francisco to San Diego,

Mail being brought over to USS *Hornet* (CVA-12) at sea near Formosa. The view is from the flight deck behind the island, September 16, 1954.

USS *Hornet* (CVA-12) refueling from USS *Passumpsic* (AO-107) while USS *Southerland* (DDR-743) refuels from the other side while in the South China Sea, September 12, 1954.

training pilots on carrier air operations. On May 4, the *Hornet* left for Pearl Harbor and Japan to begin operations. One of the big missions for the *Hornet* on this tour included participating in "Operation Passage to Freedom" to help transport over 300,000 Vietnamese soldiers, civilians and other members of the French Army from North Vietnam to South Vietnam per the Geneva Accords of 1954. This included a mobilization of over 100 US Navy ships mostly moving personnel and equipment from Haiphong to Saigon.

Capt. Norman Axtell Campbell took over the ship in July 1955. Campbell was born in 1907 and graduated from the Naval Academy in 1930. He served on the USS *Nevada* (BB-36) before going to flight school, where he graduated in 1935. He was assigned to VF-6 on the USS *Saratoga* (CV-3), and served on the USS *Minneapolis* (CA-36) leading up to World War II. During the war, he first served on the USS *Yorktown* (CV-5) until sunk, participating in the Battle of the Coral Sea and Midway among others. He was reassigned to the USS *Saratoga* (CV-3) until getting assigned as XO for the USS *Franklin D Roosevelt* (CV-42). He also served as the commander of various air groups before taking over the *Hornet.*

The rest of the time was spent on training and readiness until arriving back in San Diego in December 10, 1955. In January 1956, the *Hornet* was again designated for a refit (SCB-125) and sent to Bremerton, Washington. This ended a short two years of service where the *Hornet* did not participate in any declared conflicts but was right in the middle of some of the major Cold War incidents in the Pacific.

Anti-aircraft gunnery practice at sea near the Philippine islands, September 29, 1954.

F9F-6 Cougars taking off as next set taxi forward to be attached to the catapults near the Philippine Islands, VF-91, October 1, 1954.

The 30,000th landing on USS *Hornet* (CVA-12) was by Lt. Carl E. Smith in his F9F-5 Panther. He is congratulated by Lt. Richard Fuller (FDO), October 1, 1954.

The sun shines on two F2H-3 Banshees (VC-3) on the flight deck in the South China Sea, October 1, 1954.

Above: Four AD-6 Skyraiders from VA-95 in perfect formation over USS *Hornet* (CVA-12) in the South China Sea, October 3, 1954.

Left: Two F2H-3 Banshees (VC-3) fly over USS *Hornet* (CVA-12) in the South China Sea, October 3, 1954.

SCB-125

The next upgrade for the *Hornet* was the SCB-125 upgrade. Although this upgrade was going to be simpler, it resulted in a much more noticeable change to the look to the ship. The largest change was the new angled flight deck, a British invention. The British had been experimenting for some time with the angled flight deck idea to speed up flight operations and increase safety. The idea was that planes could land on the angled portion at the same time that planes could take off from the straight portion in the front of the ship. It also allowed planes to be stored in between the two areas while flight operations were happening. It also required less arresting wires because if the planes missed the wires it could just take back off without hitting other aircraft on the flight deck. At first, the US Navy was not too interested in the idea but eventually they came around and realized the benefits. So what was in the SCB-125 upgrade?

1. Redesign of the flight deck to include the angled portion on the stern.
2. Installation of the new Mk. 7 arresting gear system, with the number of wires reduced to five (although only four were usually used). The nylon crash barriers were also strengthened.
3. The addition of the mirror landing system (this system allowed pilots to focus on a light that was shone onto a mirror in relation to lights along the side to guide the landing).
4. The number three elevator was moved from the centerline to the starboard edge of the flight deck and was foldable.
5. Primary flight control was moved to the aft portion of the island in a position where it could easily monitor landings and take-offs.
6. A "hurricane bow" was installed, enclosing the front of the ship to strengthen the flight deck even more. This was a long overdue change from the typhoon that damaged the flight deck back in World War II.
7. Finally, air conditioning was added to many compartments and the island was better soundproofed.

This whole upgrade only took about eight months, finishing up in August 1956. On August 17, with a new commander, Capt. William Wallace Hollister, the *Hornet* began the shakedown cruise and a few months of testing out the changes to the ship. Hollister graduated from the Naval Academy in 1931 and was assigned to the USS *Nevada* (BB-36). He earned his aviator wings in 1933 and finished a degree in aeronautical engineering in 1941. During World War II, he spent two years hunting submarines in the Atlantic from shore-based Hudson and Ventura aircraft. He also spent two years in the Navy Bureau of Aeronautics as a torpedo plane design officer. He served as executive officer of the USS *Belleau Wood* (CVL-24) and USS *Boxer* (CV-21) and then commander of the USS *Kenneth Whiting* (AV-14) before taking over on the *Hornet*.

Lt. (jg) Charles P. Baker Jr. had to parachute from his disabled F9F06 Cougar over the South China Sea and was picked up by a HUP-2 Retriever from USS *Helena* (CA-75) and brought back to USS *Hornet* (CVA-12), November 7, 1954.

USS *Hornet* (CVA-12) in Puget Sound, Washington, after completing the SCB-125 refit. Changes that can be seen in this picture is the angled flight deck and the new, enclosed hurricane bow. A strange detail is the cars on the flight deck (of families moving to San Diego), August 17, 1956.

Carrier qualifications were undertaken with several air groups on the new angled flight deck. Finally in January 1957, the *Hornet* was deployed back to the Pacific with Air Group Fourteen embarked for another six-month cruise in Seventh Fleet. In May 1957, the *Hornet* suffered a damaged propeller and shaft that had to be repaired in Yokosuka, Japan, for two weeks. Historians and older naval personnel would note the irony of being repaired in the dry dock of a naval base once bombed by the same ship. Back in service, the *Hornet* was near China in June when four of her A-1 Skyraiders flew over the coast of the People's Republic of China and Chinese anti-aircraft gunners shot at the aircraft and damaged one. By July, the *Hornet* was back in San Diego, and on August 12, 1957, Capt. Thomas Francis Connolly took over the ship. For the rest of the year, the *Hornet* trained off the coast of California.

Capt. Connolly served in the navy for thirty-eight years and retired as a vice admiral. He graduated from the Naval Academy in 1933 and then went on to earn his wings immediately after. He earned a post-graduate degree in aeronautical engineering in 1942. He commanded Patrol Squadron 13 in 1943 and 1944 (seeing action in the Gilbert Islands, Marshall Islands, and over Wake Island). He was then involved in test piloting, first taking the training and serving as a test pilot and later in charge of the school. Most notably, he helped to develop the F-14 Tomcat that was partially named in his honor.

In January 1958, the *Hornet* was sent back to the Pacific again for her fourth Pacific cruise, this time with Air Group Four embarked. The *Hornet* participated

USS *Hornet* (CVA-12) in San Diego with an unusual collection of aircraft on deck including F9F-5 Panthers, an AD Skyraider, a TBF/M Avenger, and an F7U Cutlass, October 1, 1956.

in the annual "Black Ship Festival" in Japan, commemorating the arrival of Commodore Matthew C. Perry in 1853. Black Ships refers to the Japanese term for foreign ships that had been excluded from visiting Japan before then. After this cruise, the *Hornet* arrived back in Alameda on June 30, 1958 in time to participate in 4th of July celebrations. This timing had the *Hornet* missing the next major PRC incident with Taiwan that lasted from August to October 1958. The Seventh Fleet squared off against PRC aggression again and eventually the PRC backed down.

Just before arriving back in Alameda, on June 27, 1958, the *Hornet* was redesignated CVS-12 (S for anti-submarine) and ordered to Bremerton, Washington, for another refit. This would be the third major upgrade in this decade. On August 13, the *Hornet* arrived at Bremerton to begin the conversion. Capt. Marshall William White took over command of the *Hornet* on August 25, 1958. White was born in 1911 and was commissioned into the navy in 1930. In World War II, he flew air missions in the south Pacific. He graduated from the Industrial College in 1949 and then served in various command positions in air wings, including in the Korean War on the USS *Philippine Sea* (CVA-47). He is perhaps best known for his role in the USS *Pueblo Bay* (AGER-2) Inquiry in 1969.

CVS-12 Conversion

The conversion from CVA-12 to CVS-12 only took four months. The modifications were not as significant as the SCB upgrades but definitely changed the trajectory of the *Hornet* for the rest of her career as an active navy ship. Because of advances in submarine technology (by the Germans near the end of World War II and by both the Soviet and US navies after World War II), it was realized that the large carriers were vulnerable to submarine attacks. The new class of anti-submarine carriers was to be designed to protect their larger, newer brethren from submarine attacks. In addition, their missions would include defense of coasts and island bases by preventing enemy submarines from getting close enough to launch their munitions. As with other weapon systems, there was also an element of deterrence in the new ships. By proving to the Soviets that their submarines could regularly be tracked, they would be discouraged from trying first-strike type attacks.

The major changes to make the *Hornet* an anti-submarine carrier were the addition of large numbers of helicopters and fixed wing anti-submarine aircraft to the ship's air groups and the facilitation of anti-submarine equipment such as sonobuoys and the Mk. 44 (a light weight ship or air launched anti-submarine torpedo). The air groups now consisted primarily of S2-F Trackers and HSS-1 Seabat helicopters. This was to be the general configuration of aircraft for the rest of the *Hornet*'s career, although the models of S-2 and the types of helicopters (SeaKings were carried later) changed and they also included a few F2H Banshees and later A-4 Skyhawks for limited attack and fighter capabilities. There were also small VAW (early warning) squadrons of aircraft from the AD-5W Skyraider to the E1 Tracer.

Starting in 1959, sophisticated new anti-submarine tactics were created. Typically, a CVS would have four S-2s in the air on submarine patrol along with one AEW (airborne early warning) aircraft. Four helicopters were on standby to close in on submarines the S-2s found with their magnetic anomaly detectors (MAD) and sonobuoys. In addition, there were typically eight destroyers with the carrier: four on close anti-submarine support and four others at longer distances tracking submarines. The *Hornet* conducted extensive ASW training as part of its April to November West Pacific (WestPac) Cruise in 1959. During this time, the *Hornet* recorded its 1,000th helicopter and 53,000th fixed wing aircraft landing. Also in July–August 1959, the *Hornet* hosted the midshipmen from the naval academy.

During this time, the *Hornet*'s long association with NASA began when Malcom Scott Carpenter was chosen to be a Mercury 7 astronaut. Carpenter had served as the air intelligence officer aboard the *Hornet* in 1958. He backed up John Glenn on the Mercury-Atlas 6 mission and he was the pilot for the Mercury-Atlas 7 mission. This made Carpenter the second American to orbit the Earth and the fourth American in space. He spent a total of four hours and fifty-six minutes in space.

Capt. Ernest Edward Christensen took over in November 1959 and was in command when the *Hornet* was on its March–November 1960 WestPac Cruise.

USS *Hornet* (CVA-12) in Pearl Harbor with a beautiful view of the hills behind, January 26, 1957.

Of note during this cruise was the switch from Banshees to Skyhawks. Christensen graduated from the Naval Academy in 1934 and completed flight training in 1937. In his early career he served on the USS *Idaho* (BB-24), USS *Saratoga* (CV-3), and USS *Nevada* (BB-36). He served in the Bureau of Ordnance at various times in his career. In World War II, he commanded the USS *Rehoboth* (AVP-50) in 1944 and 1945. Before taking over the *Hornet*, he served as executive officer of the USS *Pine Island* (AV-12) and as operations officer of the USS *North Sound* (AVM-1).

Capt. David Charles Richardson took command in November 1960. He graduated from the Naval Academy in 1936, served two years on USS *Tennessee* (BB-43) and then entered flight training, earning his wings in 1940. During World War II, he served on the USS *Saratoga* (CV-3) and on Guadalcanal, shooting down four aircraft. He commanded Air Group Thirteen during the Korean War and later commanded the USS *Cimarron* (AO-22) before taking over the *Hornet*.

Richardson oversaw a four-month overhaul in Bremerton that started in February 1961. The squadron configuration during this time was two ten-plane S-2 Tracker squadrons, sixteen Seabat helicopters (later SeaKings) and 2 E-1 Tracer AEW aircraft. As mentioned before, at various times there were four A-4 Skyhawks.

Capt. Hoyt Dobbs Mann took command in October 1961. Mann graduated from the Naval Academy in 1936 and was assigned to the USS *Astoria* (CA-34) before taking flight training and earning his wings in 1939. He was known for the fact that he served in scout squadron two on the USS *Lexington* (CV-2) during the Battle of Coral Sea when she was sunk. He was the commander of USS *Firedrake* (AE-14) before taking over the *Hornet*.

The major event of note during Mann's captaincy was the crew of the *Hornet* participating in fighting the Hollywood Hills fire of November 1961 and supplying electricity to the city. The *Hornet* sent fifty-four of her damage control crew to help fight the fire and later many more crew to help in the cleanup.

In June 1962, the *Hornet* left for her seventh West Pacific (WestPac) cruise and in September Captain Ellis Jay Fisher took over command. Fisher, who graduated from the Naval Academy in 1939, was awarded the Navy Cross in World War II for actions in the Bismarck Sea as commander of VP-34 (flying PBY-5s) in late 1943 to early 1944. In multiple night attacks on a Japanese convoy, he was credited with sinking seven enemy vessels, probably sinking two others, and damage to three more.

Four of the eight 5-inch guns were removed at Long Beach during a short overhaul in this time. By this time, Air Group 57 was on the ship and they had all the latest weaponry including electronic rockets, homing torpedoes, depth charges, and even nuclear depth bombs. At this time, the Mk. 44 torpedo was starting to be replaced by the better Mk. 46.

In September 1963, Capt. John Ingolf Hardy was placed in command and the *Hornet* left for her eighth WestPac cruise in October. Hardy graduated from the Naval Academy in 1940, served on the USS *Pensacola* (CA-24) and then earned his wings in 1943. He later served as the executive officer of Patrol Squadron One, commanding officer of Patrol Squadron 22, executive officer of the USS *Princeton* (CV-37), and finally commanding officer of the USS *Taluga* (AO-62) before taking over the *Hornet*. In between, he finished courses at the Industrial College of the Armed Forces.

During this cruise, a tragic accident occurred when an AD-5W aircraft crashed into the flight deck, causing a large explosion. Several crewmembers were killed and wounded. The damage was so severe that the other aircraft that were still aloft were ordered to land at Atsugi NAS and two were lost (with their crews) when they ran out of fuel.

Also during this cruise the *Hornet* participated in Operation Gold Ball, an exercise witnessed by John F. Kennedy himself just a few months before his assassination. Later in the year, the *Hornet* participated in Operation Backpack, a joint exercise with Nationalist China highlighting amphibious and anti-submarine operations.

Another tragic accident occurred on February 16, 1964 when an A3J Vigilante crashed on the *Hornet*. The barricade was not able to be raised in time and the pilot was killed and two other crewmen were injured. After returning to Long Beach in April 1964, the *Hornet* traveled to San Francisco in June 1964 to prepare for the FRAM II conversion.

FRAM II Conversion

As the US Navy moved into the mid-1960s, it faced some big decisions. The huge fleet from World War II was really showing its age. Should the life of these ships be extended by further modernization or should they be scrapped in favor of new ships? The FRAM (Fleet Rehabilitation and Modernization) program was the navy's attempt to get the best use out of the old ships. The first part focused on destroyers, but during FRAM II, the navy evaluated their *Essex*-class carriers (among other ship types). They had to first determine if the old ships were still structurally sound and if so, modernize them. The *Hornet* began the upgrade in July 1964 and it was complete in February 1965.

Capt. Mayo Addison Hadden took over command just as this update was about to start. Hadden was born in 1916 and enlisted in the naval aviation program in 1942. He was first deployed to naval air stations but was eventually transferred to the Pacific where he became an ace with eight confirmed enemy aircraft destroyed. He won two Silver Stars and two Distinguished Service Crosses for his actions in the Pacific. He became the navigator for the USS *Yorktown* (CV-10), and then was the commanding officer of the USS *Graffias* (AF-29) before taking over the *Hornet.*

The main components of FRAM II for the *Hornet* were:

1. A complete overhaul of the boilers and propulsion gear systems.
2. An improved replenishment rig (for at sea refueling).
3. A full upgrade to the sonar system including the Installation of a SQS-23 sonar set in the bow dome. The idea being that the *Hornet* had to be able to detect submarines for itself considering its anti-submarine role would be bringing it into closer contact with the enemy. This also forced the movement of the port side anchor to the bow so it would not interfere with the SQS-23. This sonar set was the same type that was added to the FRAM destroyers.
4. Redesign and update of the Combat Information Center to process all the new anti-submarine data, including an aircraft data link.
5. Installation of a closed circuit television system.
6. Addition of aluminum planking on the flight deck landing area to help reinforce it for heavier aircraft operations.
7. Installation of GEMINI space recovery facilities in the Communications department.
8. Upgrade to the new Fresenel Lens Landing System also known as PLAT (Pilot Landing Aid, Television).
9. At the same time, aircraft updates included the new E-1B "Tracer" (a modified S-2), and SH-3a SeaKing helicopters.

These modifications turned the *Hornet* into a highly sophisticated anti-submarine vessel. With its new ASW Ship Command and Control Systems (ASWSCCS) and Anti-submarine Contact Analysis Center (ASCAC), the *Hornet* was now part of an advanced system linking data between aircraft, escorting destroyers and even Allied submarines into an efficient submarine hunting machine. At this time there were over 400 submarines in the Russian fleet so the need for anti-submarine forces was very important.

The *Hornet* began testing and training the crew with their newly upgraded ship. One of the new SeaKings set a new helicopter endurance record by flying from the *Hornet* to the USS *Franklin D. Roosevelt* (CV-42) for sixteen hours and over 2,000 miles in March 1965. Once this was complete, the *Hornet* returned to Long Beach on March 26. Capt. William McKnight Pardee took command on July 1, 1965, and Air Group 57 was embarked. Pardee was born in 1920 and graduated from the Naval Academy in 1943. During World War II, he served on destroyers in the Pacific. In 1948, he earned his wings as a naval aviator and served in various ASW squadrons (VS-21, VS-81, and VS-37) before being named as commander of USS *Mispillion* (AO-105) and then finally taking over the *Hornet*. On August 11, 1965, the *Hornet* left port for her ninth WestPac cruise and her first Vietnam War cruise.

In the twenty years from the end of World War II until the *Hornet*'s first Vietnam cruise, the ship underwent three major upgrades as well as many minor overhauls. All of the changes needed to go from handling primarily piston-engined aircraft to the newer, heavier jet aircraft were made. Redundant systems such as the old 20-mm anti-aircraft guns were removed. Major changes like adding an angled flight deck were made. The island was completely redesigned. Elevators were moved around. Provisions for ASW work were made, including extensive upgrades to equipment needed to track submarines and analyze the data. Helicopters became an important part of the ship's air groups. The basic frame of the *Essex*-class carrier was able to accept all these changes and continue to flourish in a navy alongside brand-new ships.

5

The Vietnam War and the Moon Landings

The Vietnam War

Ever since the French had been defeated in the fighting in Vietnam, the US had taken a larger interest in the area. The US was bound by treaty to resist the spread of Communism in the Pacific area. This was somewhat easier on islands where the Seventh Fleet could decisively intervene, such as in the Philippines or Taiwan. It was more difficult on the mainland because it was harder to interdict the flow of weapons and troops to these areas. Vietnam was a perfect example. Since Vietnam shared a border with China, it was easier for both Chinese and Soviet weapons and advisors to get to the North Vietnamese.

Tensions rose between North and South Vietnam, and the US increased its involvement in an attempt to prop up the struggling South Vietnamese regime. Starting in the early 1960s, US advisors, Special Forces, support units, and other specialists began to arrive in South Vietnam. The situation did not improve and it was becoming more clear that the US either needed a full commitment or to let South Vietnam fall. On August 2, 1964, USS *Maddox* (DD-731) was attacked by North Vietnamese Navy patrol boats. A second incident two days later involving the *Maddox* and another destroyer was later determined to have not occurred. There was evidence that it was known this second battle was not a real one, but this was suppressed to make the case stronger against North Vietnam. The US Navy retaliated by using carrier aircraft to bomb North Vietnamese naval facilities. President Johnson drafted the Gulf of Tonkin Resolution (signed into law on August 11, 1964) that stopped short of declaring war, but announced that the US would do everything else needed to support South Vietnam, including heavy commitment of the military.

The US Navy began operating in two different stations off the coast of Vietnam: Yankee Station in the Gulf of Tonkin in the north and Dixie Station near the Mekong Delta in the south. The missions of the navy were many: ground support for troops, strategic bombing, supply interdiction (both on land and at sea),

Loading a rocket pod an on S-2F Tracker on USS *Hornet* (CVS-12) at Yankee Station off the coast of North Vietnam, June 1967.

S2-F Trackers preparing for take-off from USS *Hornet* (CVS-12) at Yankee Station off the coast of North Vietnam. Note the radio antennas off the front of the flight deck are in their down position for the launching of aircraft, September 5, 1967.

intelligence gathering, search and rescue, and more. In addition to the *Hornet*, the USS *Yorktown* (CVS-10), USS *Bennington* (CVS-20), and the USS *Kearsarge* (CVS-33) would all rotate through Vietnam from 1965–1969. The *Hornet* would end up serving three tours off of Vietnam and was involved in many of the missions listed above. Although classified as an ASW carrier, the *Hornet* would be called on heavily for other missions as well.

The Vietnam War would end up as a testing ground for Cold War technologies and tactics. Both side would continuously upgrade their technology and modify their tactics as the conflict carried on. For the navy, the conflict would test the durability of the fleet in many ways. The navy committed to having three carriers on station at all times and this strained the Pacific (Seventh) Fleet and ended up straining the other fleets as well. Ships were transferred in from the other fleets to fill the gaps, rotations were extended, maintenance was delayed, aviators were lost (and had to be replaced), and morale was impacted.

When it came to aerial missions, the navy had several advantages over the air force and this showed over the course of the war. First of all, navy pilots had to be better trained because the not so simple exercise of landing on a moving carrier deck required more skill than landing on a straight runway. The fleet at Yankee Station was closer to their targets than the air force bases so navy pilots had to spend less time over hostile territory and the North Vietnamese had less time to prepare their defenses before strikes occurred. The navy aircraft could often fly under the radar and get even closer before detected. The ships of the fleet could aid their aircraft with shipboard radar as well as air to air missiles on occasion.

Both the air force and the navy used the F-4 Phantom as a fighter aircraft (among other roles), but the navy also used the F-8 Crusader. The F-8 was the last of the so-called cannon fighters and their training reflected that. They had missiles and did use them but they were also used to dog-fighting. The navy training emphasized a loose formation of two aircraft and strikes with one missile at a time (the air force training tended to teach firing large salvoes of missiles) meaning their fighters could often fight much longer. The performance of missiles during the war was mixed. Far too many times the missiles would not launch, would not arm properly, or would lose their target after launch. The F-4 Phantom did not have cannons to start the conflict, but they were added later in recognition of the disadvantages of not having that option. Their main opponents were MiG-17s, MiG-19s, and MiG-21s. Kill ratios were varied and not always in the favor of the US aircraft.

Over the course of the war, the navy F-8 pilots consistently had the best kill ratio. The navy studied the effectiveness of their aircraft and concluded that they needed more training. Thus the very successful Top Gun program was born. There was some exchange of tactics between air force pilots and navy pilots, and this was helpful for the air force pilots. However, the strain on the carrier pilots was very high, with over 800 killed or captured during the Vietnam War.

The ASW carriers such as the *Hornet* were not as heavily involved in air-to-air combat over North Vietnam since they had very few strike aircraft and no fighters. However, they were heavily involved in search and rescue work and that was often dangerous, conducted under fire and over enemy territory. They were also impacted by the extended deployments and high demand for carrier pilots.

On August 19, 1965, the *Hornet* arrived in Pearl Harbor for an extensive Operational Readiness Evaluation (ORE). After passing this, the crew was given two weeks of rest and relaxation in Hawaii before heading out to Japan and on their way to the South China Sea and the Vietnam War. On September 20, 1965, the USS *Epperson* (DD-719) collided with the *Hornet*, although both ships only suffered minor damage. Finally, on October 4, the *Hornet* left Yokosuka and started intense around-the-clock search and rescue operations. The *Hornet*'s A-4E Skyraiders were detached to the *USS Midway (CVA-41)* where they conducted over 100 combat missions. For the rest of the year, the *Hornet* moved between Yankee Station, Subic Bay, Hong Kong, and the Sea of Japan for a mix of combat duty, exercises, and time off. After losing two aircraft overboard during extremely bad weather near Korea, the *Hornet* was back in Subic Bay for Christmas 1965.

At the beginning of 1966, the *Hornet* was back in the Gulf of Tonkin. On January 22, 1966, an S-2 Tracker with its crew (William S. Forman, Erwin B. Templin Jr., Edmund H. Frenyea, and Robert S. Sennett) was lost. Lt. Walter R. Davis, who was on duty in the CIC when this incident happened, said:

> I was the aircraft controller on watch In CIC on the night of January 22, 1966 for the ill-fated mission of six S2F's. I was an Ensign at that time and my watch assignment was to track each of the aircraft on the radar repeater (using the radar returns from the SPN-43) while maintaining voice radio contact with each aircraft.
>
> The *Hornet*'s radio call sign was "Judo" & each plane from VS-35 was "Sungod" followed by their aircraft number. "Sungod twelve this is Judo, over." I vectored each plane out to its assigned search area and monitored location & activity. Each of the six planes showed as a solid blip on my radar repeater and I periodically asked each aircraft to squawk their individual IFF code to double check identity.
>
> Then, midway through the mission, the radar blip from Bill Forman's Sungod twelve plane suddenly disappeared from the scope and was non-responsive to my voice radio calls. "Sungod twelve, this is Judo, come in, over." We never heard from Bill Foreman's plane again! *Hornet* and its squadrons (VS-35; VS-37; and HS-2) moved northward from Yankee Station into the Tonkin Gulf to search round the clock for Sungod twelve for several days before the search was terminated. Nothing was found.

After some searching, a few items were found: an empty life raft and a helmet that seemed to be from one of the crew. There were different rumors about what happened and the North Vietnamese did claim to have shot down an aircraft that day. There was some speculation that at least one of the crew were taken captive but none of them were released after prisoners were exchanged following the war. To this day, they are still listed as missing in action.

The *Hornet* was pulled away from the area to monitor a Soviet submarine that was sighted near Taiwan and the Philippines. After returning to Japan, the *Hornet*, was ordered down to Australia for a short joint exercise.

On February 17, 1966, the *Hornet* began the long trip to Australia. Along the way, the ship passed by many of the famous battlefields of World War II. This included Iwo Jima, Saipan, Truk, Guam, and Guadalcanal. The crew paid tribute as they passed each battle site but they paid their largest respects when they reached the Santa Cruz Islands and the approximate location where the USS *Hornet* CV-8 was sunk almost twenty-four years earlier.

The Battle of the Santa Cruz Islands was one of many fierce naval battles during the Guadalcanal campaign in World War II. Fought on October 25–26, 1942, the battle was a tactical Japanese victory, but in reality, it was a strategic Japanese defeat. The US Navy lost the USS *Hornet* (CV-8), USS *Porter* (DD-356), and suffered damage to the USS *Enterprise* (CV-6), USS *South Dakota* (BB-57), USS *San Juan* (CL-54), and USS *Smith* (DD-378). The US lost eighty-one aircraft and twenty-six airmen. The Japanese had no ships sunk but suffered damage to the *CV Shōkaku, CVL Zuihō, CA Chikuma*, and a destroyer. Most importantly, they lost ninety-nine aircraft and 148 irreplaceable airmen. This meant that out of over 700 airmen that had participated in the Pearl Harbor attack, they had already lost over 400. Soon after this, the Allies were able to win on the ground and with surface forces in the waters around Guadalcanal and the Japanese lost the campaign.

The *Hornet* reached Australia on March 1, 1966 and conducted a thirty-six-hour exercise with their counterparts in the Australian Navy. When this was complete the tired crew headed home, reaching Long Beach on March 23.

Capt. Van Vernon Eason assumed command of the *Hornet* on April 1, 1966. Eason was born in 1919 and joined the navy in 1940. He graduated with his wings as a naval aviator in 1941. He was a flight instructor at Corpus Cristi NAS until 1943. He joined VT-10 on the USS *Enterprise* (CV-6) as the squadron executive officer. He led the first night carrier strike against Truk and fought in the Battle of the Philippine Sea where his plane was damaged and he ended up having to ditch and was later picked up by a destroyer. He won a Navy Cross and a Distinguished Flying Cross. Later he served in various roles until given command of the USS *Paricutin* (AE-18) before taking over on the *Hornet.*

After heading back to Hawaii, the *Hornet* hosted 300 midshipmen for almost three weeks before heading back to Alameda. Just like when they had done this before, this was an opportunity for the midshipmen to learn what it was like to operate a ship in real conditions. On July 4, 1966, the *Hornet* had an open house on board and over 9,000 guests visited.

In August 1966, the *Hornet* sailed for Pearl Harbor but on a different mission. For the first time, the *Hornet* was directly participating in the Apollo program. The AS-202 mission used CSM-011, a block 1 command module, to conduct an un-crewed, suborbital test flight. The crew of the *Hornet* practiced the recovery process before heading out to rendezvous with the returning module. CSM-011 was launched with a Saturn IB rocket and was used to test both the rocket and

USS *Hornet* (CVS-12) underway in the Pacific, 1968.

the navigation and fuel cells of the command module. The entire mission lasted one hour, thirty-three minutes, and two seconds. CM-011 splashed down over 205 nautical miles from where it was planned to land, and it took the *Hornet* over eight hours to reach it southeast of Wake Island. Note the switch from CSM-011 to CM-011 since the service module was no longer with the command module. Once on site, the recovery process went smoothly, and the *Hornet* dropped off the module in Long Beach on September 2, 1966. This module is currently on display at the USS *Hornet* museum.

After this recovery, the ship and its crew were given a long, well-deserved rest. The ship's status was changed to restricted availability/holiday leave and this lasted for the rest of the year. The year 1967 started with training and exercises. Air Group 57 was embarked and their first exercise was called "Snatch Block," which started on January 5, 1967. A snatch block is a heavy duty pulley in a metal casing. The exercise was a combined surface, ASW, anti-aircraft, and air strike exercise. This was followed by ten days of ASW training in February. Capt. Gordon Hubbard Robertson took over command on February 27. Robertson was commissioned as a naval aviator in March 1943 at Pensacola, Florida, and flew Helldivers in Air Group Two from the *Hornet* in World War II before going on to serve in the Korean and Vietnam Wars. His unit was awarded the Presidential Unit Citation during its service from March to October 1944 and he personally was awarded an Air Medal and a Gold Star *in lieu* of a second air medal for his service in World War II.

On March 27, 1967, the *Hornet* and crew left for their tenth WestPac cruise. They were diverted to track a Russian submarine in the central Pacific. The *Hornet* was successful and tracked the submarine until it left the area. In May, they had a joint exercise with the Japanese and Korean Navies in the Sea of Japan. For the entire exercise, they were shadowed by Soviet surface and air forces. For the rest of May and June, the *Hornet* was back at Yankee Station tracking ship movements. During this period, the *Hornet* recorded the 100,000th aircraft landing. In July, the *Hornet* participated in another joint operation, this time with the Navies of SEATO. In August, they were back in Hong Kong before a four-week stint at Yankee Station from September and into October before returning to Long beach on October 28, 1967.

From the end of 1967 until May 1968, the *Hornet* underwent a long overhaul in Long Beach Naval Shipyard. During this time, the ship was checked from top to bottom and every worn out piece of equipment was replaced, every piece of rust scrubbed off and every surface repainted. It had been a heavy year for the *Hornet* with over 10,000 aircraft landings this year alone.

Capt. Jackson A. Stockton took over command on February 23, 1968. Stockton graduated from the Naval Academy in 1943 and initially served on the USS *Denver* (CL-58) in the Pacific. He opted for flight training and completed that in late 1945. He specialized in ASW and commanded VS-39, CVSG-53, and then USS *Elokomin* (AO-55) before taking over the *Hornet*.

The *Hornet* then went out on a fast cruise to initiate new sailors and re-familiarize the veterans. This was followed by six weeks of training, carrier

USS *Hornet* (CVS-12) ready room, 1968.

qualifications, and ASW operations. After finishing the ORE, the *Hornet* was back at Yokosuka to start their eleventh WestPac cruise and another assignment to Yankee Station in October. Instead they were sent to the south and conducted ASW and search and rescue operations there. The balance of the year was spent back and forth between Hong Kong, Japan, and Vietnam.

Bruce Douglas was an interior communications electrician on the *Hornet* from 1968–1970. He is a docent at the *Hornet* museum these days. He was responsible for anything that needed to be communicated inside the ship. This included intercoms, telephones, paging systems, sound powered telephones, gyrocompasses, etc.

When first assigned to the ship, Douglas was assigned to the enlisted galley. New crew members typically get assigned to these disliked jobs since each department had to supply a certain amount of man-hours to jobs like working in the galley. For Douglas this was his favorite job he ever had in the navy. He was a "Reefer Rat" and that meant he was one of the sailors that had to go down three decks to the sixth deck where the refrigerators were to bring supplies up to the galley. Everything had to be hand carried up those three decks in large canvas sacks. For fun, they made it a competition to see who could carry the most or go the fastest.

There were perks to working as a Reefer Rat. If a tub of ice cream was dropped and broke, there was always a batch of strawberries that got dropped too! This was by accident of course. They also had access to the leftovers from the admiral and captain's mess and while the food was the same it was always prepared better.

Bruce Douglas on the USS *Hornet* Museum today.

Bruce Douglas on USS *Hornet* (CVS-12), 1969.

Marine Joe Holt with President Nixon on USS *Hornet* (CVS-12), July 24, 1969.

Joe Holt was a Marine in the Marine Detachment (MARDET) during 1968 and 1969. After serving a tour in Vietnam in 1966–1967 as a rifleman, he was moved over to the *Hornet*. Joe is now a volunteer on the *Hornet* and was significantly involved in the beginning of *Hornet*'s life as a museum. He has lots of stories to tell, mostly about how bored he and the other Marines on board were, especially those that had served on the ground in Vietnam. One of his stories was about his one experience as a passenger on an S-2 flight:

> Sometimes my sea-stories don't end up the way I'd like them. This particular story I've told for forty years. You'd think I'd change the ending by this time.
>
> I'm lucky to have my old ship, the USS *Hornet*, docked somewhat near me in Alameda, California. Every so often I take friends or friends of friends up and give'em a "Joe tour" of the ship, regaling them with my yarns, mostly factual. They seem to appreciate it, hearing things that really happened exactly where they were standing at the time.
>
> One story I rarely tell, simply because most civilians wouldn't get the point, is the time I had the opportunity to be launched off the ship in an S-2F, one of the ship's anti-submarine aircraft. Everybody pretty much agrees being catapulted off a ship is pretty exciting, but unless you've actually experienced it, it's hard to imagine.
>
> The morning we'd pulled out of Singapore one of my guys in the Detachment had mentioned in passing it was possible to get a seat on regular patrol flights. He'd actually done it on a previous cruise.
>
> An S-2 was a four-seater airplane; the four being the Pilot, Co-pilot, Techy guy (Geez, how am I supposed to know what he's called?), and one other seat for some other techy guy on the right side of the airplane. That extra seat, where the Navy didn't seem to need a guy, was the one any lucky passenger would sit in.
>
> Now it must be pointed out that Christiansen was the guy who told me about this treat. PFC Christiansen. He and I hadn't seen eye to eye on a lot of things those months previous, so at first I didn't really believe this could actually be true, him being exactly the kind of snuffy who'd play a prank on his Corporal. Play a prank? How polite is that?
>
> This wariness of Christiansen, together with my natural suspicion of sailors, led me down the passageway toward the Pilot's Ready Rooms with only the slightest hope of actually finding someone who not only could confirm the possibility, but tell me the procedure if such could be the case.
>
> On the cruise I'd passed the Ready Rooms often, but I'd never actually had business with any of the air group guys, so as I approached the entrance I was delighted a Petty Officer was coming out, a convenient guy to ask.
>
> Me, trying to maintain my Marine Corps demeanor of superiority, casually asked this sailor if it was true; could a fella' get a ride in an S-2? "Sure," he said, "We've got a flight at eleven. Why don't you go back to the parachute locker and draw a jump suit and helmet and meet me back here at ten thirty?" He then proceeded down the passageway leaving me standing there in the passageway still thinking this was some sort of conspiracy. Wow, that was easy, but who ever heard of a parachute locker?

I went up to the Hangar Bay and I walked up to the first sailor I saw who was working on an airplane (The Hangar Bay was full of S-2s, our air group) and asked, "If there was such a thing as a parachute locker, where would I find it?" He didn't seem to think that was a silly question; that was a start. He then directed me to said parachute locker toward the rear of the Hangar Bay. I walked in, told 'em what I was doing, and in no time at all, the sailor handed me a green jump suit and a helmet. The helmet was extremely cool, white with one of those pull down arrangements to keep the sun out of your eyes.

I took this stuff back to my compartment and started to change into it. The other guys in the compartment were as entertained as could be with me and my new outfit. I even put the helmet on to see what I looked like in the mirror. Soooo cool! (I wish I would have thought to have somebody take my picture.) I was immediately christened 'Joe Jet' by the guys.

At ten-thirty sharp I met up with the sailor in front of the Ready Room. We went directly up to the flight deck and toward an S-2 that was lined up on the Port catapult. He started explaining to me his duties prior to launch, all kinds of last minute checks just to see they'd been properly completed. He even kicked the tires, but smiled when he did it. He climbed up on the top of the airplane and checked to see if the gas tank was full. Good idea.

He took me inside the plane, had me sit down in my seat, and adjusted my seat belt assembly to the proper tightness. Then the pilot and co-pilot boarded and gave me a brief howdy as they clambered into their seats. Somebody had brought aboard four box lunches for the flight, which I thought was a bit silly considering we were only gonna' be in the air for a few hours.

After we were all seated and situated, before the engines were fired up, the techie guy gave me some last instructions on evacuation of the airplane in case a problem should arise. We were to roll out of our seats and crawl towards the hatch behind us, then tumble out and pull the cord. That's the kinda' thing a guy remembers.

He asked me if I would give him a signal just prior to the launch. I had a window near my seat, he didn't, and there was gonna' be a flight deck guy who was going to give a signal to the pilots to take off. The sailor asked me to be kind enough to yell "Now!" when the signal was given. The techie just wanted to know so he could prepare himself for the actual launch. Simultaneously, he told me to hold my arms directly out in front of me in the rare case the radio console flew out of its mountings during the abrupt take off. A huge arrangement, with all kinds of knobs and switches directly in front of me. Got it. Then they started the engines. Very cool.

So there I sat, absolutely thrilled, engines goin' like hell, staring at the flight deck sailor waiting for a signal, with my arms out in front of me. Seconds went by, me concentrating on my instant response to the signal. As expected, the flight deck guy quickly dropped his arm. So, this was it. This was the one instant I'd been waiting for. The reality of a catapult launch was a rude awakening for sure. My mouth opened. I started to say "Now!" Honest, I did. But all I got out was "Nnnn" before all hell broke loose. The thrust was incredible! The force of the launch made my hands instantly fly to the rear, damn near hitting me in the face. Startled, I had to physically push my arms out again towards the console, at the same time completing my signal, "ow!" This only took a blink of an eye. Then we were airborne, me feeling a bit embarrassed my arms hadn't stayed outstretched as I'd been advised to do, but still completely impressed with the thrust, the violence of the take off. Wow.

The techie didn't seem to notice if I'd kept my hands out, and it took only a minute or two for me to calm down and enjoy the flight. Being less than a day out of Singapore there was a certain amount of shipping we saw below, but the marvel were the myriad of green islands scattered across the ocean, surf meeting the lush green of the islands, contrasted against the blue of the Pacific. Absolutely gorgeous.

The pilot and co-pilot opened hatches over their heads which allowed a warm, south-sea breeze waft through the airplane. The techie told me to point to any part of the console, which I did, then directed me to a particular switch which he told me to flip. Armed Forces Radio. Rolling Stones singing "Black on Black." Loud. Perfect.

At some point I opened my box lunch. Fried chicken and apple pie! What's this all about? I never saw either of these on the mess decks. Those guys in the air group sure had it swell.

So there I was flying along above the beautiful, island specked ocean, appreciating the warm, sweet breeze, listening to tunes, eating my fried chicken. Life simply didn't get any better than this.

I don't remember how long we were in the air. Two hours? Four? Anyway, when we approached the *Hornet* it was nothin' but a gray speck on the ocean, then all at once we were aboard, me staring at the superstructure. The landing was almost anticlimactic compared with the launch.

It was such a kick. I told the guys all about it, but they just couldn't imagine how extraordinary it was.

Fast forward twenty-five years. One night I was lying in bed, thinking about all sorts of stuff while I was trying to fall asleep, when it hit me like a ton of bricks.

Holy Shit! Oh, Maaaaaaaaan! Was I a moron or what? Only then, after all those years, did it finally occur to me I'd been made a fool of by the Navy.

There was no way in hell that console could "jump out of its mountings." The techie guy knew from the get-go my physical reaction to the launch, and he wanted to watch me flap around like an idiot. Maybe even, with luck, he'd get to see me hit myself in the face. I betcha [*sic*] he's been telling that story to his friends and family all these years, about the moron Marine who got a flight off the USS *Hornet.*

Starting in early 1969, the *Hornet* participated in Operation Market Time, which targeted ships transporting supplies from North Vietnam to South Vietnam. This operation was primarily the responsibility of Task Force 115, subordinate to the Seventh Fleet, and it was undertaken by large ships and aircraft out to sea, then destroyers and coast guard cutters closer in, and finally by fast patrol boats close to the shore. The *Hornet*'s responsibility was further out, providing aircraft to find and track vessels trying to move through the blockade. This operation was highly successful. During the rest of this tour they also participated in the Beacon Lamp exercise with the Australian Navy (specifically with HMAS *Perth* [DD 38], lead ship of her class and veteran of three deployments to Vietnam), and trips to Subic Bay, Singapore and Japan. The crew was granted nine days of R&R in Singapore. Finally they headed back to the west coast of the US in May, returning May 13, 1969 and four weeks of limited availability again. This was the last Vietnam tour for the *Hornet* and the ship was about to enter a new phase of its history as part of the first two Moon landings.

Apollo 11

As John F. Kennedy had promised in his speech in 1961, NASA was working diligently towards its goal of landing a man on the Moon and bringing him safely back to Earth before the end of the decade. During the decade of the 1960s steady progress was made in the space program. The US went from sending men to space, to orbiting the Earth, to orbiting the Moon, and finally to a simulated separation of the lunar module and lunar lander. The US was now ready to launch the final test mission: landing men on the Moon's surface and bringing them back to Earth as Kennedy had promised

The crew of a SH-3D SeaKing in the hangar bay of USS *Hornet* (CVS-12). Pictured are Cdr. Donald Jones and Lt. (jg) Bruce Johnson of HS-4 about to recover the Apollo 11 astronauts 900 miles southwest of Hawaii, July 24, 1969.

On May 23, 1969, Capt. Carl J. Seiberlich was given command of the *Hornet.* Seiberlich was an experienced navy veteran, having served on the USS *Mayo* (DD-422) in World War II, fighting in the battles of Salerno, Anzio, and Okinawa, while also being on station in Tokyo Bay for the signing of the surrender documents in September 1945. After the war, he trained to fly Blimps and he was responsible for several blimp milestones such as the first night landing on an aircraft carrier and the first cross country flight in a blimp. In 1952, he switched to fixed-wing aircraft, flying PB4Y-2 Privateers, P2V Neptunes, and, finally, S-2 Trackers. He commanded VS-26, flying S-2 Trackers on the USS *Randolph* (CV-15). Later in his career, in 1963 while navigator aboard the USS *Interpid* (CV-11), he qualified on HSS-1 Seabat Helicopters. This made him the first naval flyer to qualify on blimps, fixed-wing aircraft, and helicopters. Seiberlich's first command was of the USS *Salmonie* (AO-26) in 1967. After that he was given the *Hornet.*

A few days after taking command of the *Hornet*, on June 1, 1969, Seiberlich confirmed the rumors and told the crew that they had been assigned as the recovery ship for the Apollo 11 Moon mission. On June 16, the *Hornet* left Long Beach to begin a one-week training cruise for their new mission. On board they had 125 civilian and military specialists from NASA, General Electric, ABC TV, and the more. Special equipment was also brought into the hangar bay, including several trailers and communication structures. To help make space, they left most of their regular air crew at Long Beach.

The *Hornet* was an especially well-suited ship to be a recovery vehicle for the Apollo 11 crew and command module. The large hangar bay had plenty of room for extra equipment. The crew quarters were large enough to fit extra "crew" for a special mission such as this, especially considering they left most of their air group behind. Their long experience with helicopters and fixed-wing EAW aircraft was very relevant for locating and picking up the command module and the astronauts. All in all, ten *Essex*-class carriers would be involved in the various manned spacecraft programs (Mercury, Gemini, Apollo, and Skylab) including USS *Essex* (CV-9), USS *Yorktown* (CV-10), USS *Intrepid* (CV-11), USS *Hornet* (CV-12), USS *Ticonderoga* (CV-14), USS *Randolph* (CV-15), USS *Wasp* (CV-18), USS *Kearsarge* (CV-33), USS *Princeton* (CV-37), and USS *Lake Champlain* (CV-39).

On June 29, the *Hornet* left for Pearl Harbor. When the ship reached Hawaii on July 2, they picked up the rest of their special crew and equipment. This included another 200 specialists and their equipment, a full-size mockup of the command module, two modified Airstream trailers (to be used as mobile quarantine facilities or MQFs) and a modified air group that now included three C1-A Traders, four E1-B Tracers (VAW-11), and eight SH-3D SeaKings (Helo ASW 4). Helo ASW Squadron 4 had previous recovery experience so was a perfect choice for this mission.

The MQF was a very interesting piece of equipment. From the outside it could be mistaken as a simple Airstream mobile home trailer. However, Airstream and Melpar Inc. had made many modifications to their normal designs to make the MQF perform as requested by NASA. The Interagency Committee on Back

Contamination (ICBC) had been created early in the Apollo program to consider the threat of dangerous microorganisms returning from the Moon with the astronauts. Their design specifications led to the MQF. It was airtight and used a combination of fans, filters, and a pressure lock to prevent any dangerous organisms from escaping outside of the MQF. It was built with a complete communication system and multiple, redundant power systems so the astronauts could stay inside for long periods of time if needed. It was also designed to be sturdy enough to be transported easily by ship, ground transport or by air.

The plan for the recovery operation was for there to be three E1-B Tracers in the air for communication purposes and four SH-3D Seakings for the recovery. Two of the SeaKings would carry navy underwater demolition teams (UDTs) who would go into the water to assist the astronauts in decontaminating the capsule and themselves and then transferring from the capsule to the third SeaKing. The navy had five UDTs and they had picked UDT-11 for this mission. The fourth SeaKing would be on hand to photograph everything. For decontamination, the capsule would be cleaned with betadine and hypochlorite and the astronauts would put on biological isolation garments (also called BIG suits) before getting on the helicopter. Once on board the *Hornet*, the astronauts would transfer to one of the MQFs (the second was brought on board as backup in case the first one malfunctioned), where they would be quarantined.

For a week, the *Hornet* trained in the waters around Pearl Harbor. They ran a total of sixteen simulated exercises (SIMEX) in that time—more than two per day! Finally, on July 12, the *Hornet* left Pearl Harbor to head to the recovery area. At first they had to head to the abort recovery area, in case the launch of Apollo 11 failed for some reason and had to come back to Earth after one orbit. The *Hornet* was in position for this on July 14, but as we know the mission was successful so they then steamed to the originally planned final recovery spot. While on the way, they continued to practice.

While this was happening, the Apollo mission proceeded to the Moon and back. On board were three astronauts that would include the first two humans to ever step on the surface of the Moon: Tom Collins, Edwin Eugene "Buzz" Aldrin, and Neil Armstrong. Collins was an air force test pilot, Armstrong was a former navy pilot (he flew seventy-eight missions over Korea in F9F Panther), and Aldrin was an air force pilot (he flew sixty-six missions in an F-86 over Korea and shot down two MiGs). Armstrong was also famous for flying the X-15 (a joint air force/NASA experimental rocket-powered hypersonic aircraft that influenced spacecraft design). Between the three of them, they had a lot of experience in the US space program. All three had participated in the Gemini program and had been to space before Apollo 11. Collins piloted the command module around the Moon while Armstrong and Aldrin landed on the surface of the Moon and actually walked on its surface. After taking off and reconnecting with the command module, they headed back to Earth.

On July 23, the *Hornet* was informed that the expected landing site had changed and they had to steam as quickly as possible to a spot 250 miles closer to Pearl Harbor. This was due to bad weather in the original landing area. The new location was about 920 miles southwest of Pearl Harbor. The next day was the splashdown and recovery date.

The *Hornet* was very busy on July 24. In addition to launching their own aircraft, several aircraft from Hawaii were in support of the recovery mission, including two HC-130 Hercules and three Apollo range instrumentation aircraft. Also participating was the USS *Arlington* (AGMR-2), formerly the USS *Saipan* (CVL-48). The *Arlington* was a specialized communication ship and its participating in the mission was secret. Being that the first Moon landing and return was one of the most significant accomplishments of mankind to this date, the *Hornet* was visited by Marine One with President Nixon and Secretary of State Kissinger on board (after a way stop on the *Arlington* on their way from Pearl Harbor). They arrived at 5:12 a.m. to be on the *Hornet* during the recovery to greet the astronauts. Also on board were many members of the press.

The command module (Columbia) entered the atmosphere at 22,300 mph. During re-entry, the outside temperature of the module reached 5,000 degrees and there was a 200-mile flame trail behind it. During re-entry, communications were lost for over four minutes. Then the first communication between *Hornet* and Columbia occurred:

"Apollo eleven, Apollo eleven. This is *Hornet. Hornet*, over."

"Hello *Hornet*, this is Apollo eleven reading you loud and clear."

Lt. Clancey Hatteburg closes the hatch of the Apollo 11 command module after astronauts Armstrong, Collins, and Aldrin have boarded the raft, July 24, 1969.

Apollo 11 astronauts Armstrong, Collins, and Aldrin in life raft next to command module with Hatleberg, 900 miles southwest of Hawaii, July 24, 1969.

At 5:50 a.m., the UDT (swim-1) announced that Apollo 11 had splashed down safely. It was 11 miles from the *Hornet* and within a mile of its intended landing spot. When Columbia hit the water, it turned over, leaving the astronauts upside down. This was anticipated as a possibility and once the floatation bags were inflated by the astronauts on board, the craft righted itself. One team of divers went into the water and attached an extra floatation device to the module. The door was opened and the divers threw in the BIG suits, which the astronauts donned. When those were on, they came out and got into a raft to be lifted into SeaKing 66. The divers then wiped down the module with antibiotics.

The astronauts were flown back to the *Hornet* and landed on the flight deck and moved to elevator two. The helicopter was lowered to the hanger bay and pushed near the MQF. The astronauts then disembarked from the helicopter and entered the MQF along with two other new crewmembers: Dr Bill Carpentier and engineer John Hirasaki. Carpentier was there to monitor the health of the astronauts and Hirasaki was there to ensure the equipment in the MQF worked and to help with food preparation and other tasks. After a shower and a change into uniforms, they came to the window of the MQF to be greeted by President Nixon and Capt. Seiberlich. The crew made a sign that said "*Hornet* +3" in recognition of the three new crewmembers on board.

While all this was happening, the *Hornet* was steaming to the location of the command module. Once alongside, the *Hornet* used its crane to bring the

Helicopter 66 lifts one of the astronauts in its recovery basket, July 24, 1969.

Astronauts leaving Helicopter 66 on USS *Hornet* (CVS-12) to enter the MQF, July 24, 1969.

Astronauts waving as they moved towards the MQF on USS *Hornet* (CVS-12), July 24, 1969.

command module on to elevator three. This was an excellent display of seamanship to bring such a large ship next to the command module without damaging it. It was brought into the hangar bay near the MQF. A plastic tunnel was erected to connect the MQF to the command module so Hirasaki could retrieve the moon rocks and other items. The moon rocks were transferred to aircraft and flown to Hawaii on two different routes before they went to Houston. The entire recovery mission had been essentially flawless.

Joe Holt and Bruce Douglas were both on the *Hornet* during the Apollo 11 recovery. Holt remembers how proud the Marines were to have the actual president on board the ship and he managed to get captured in a few photos standing near the president during the Apollo 11 recovery. Douglas remembers that everyone was told to stay below the hangar bay unless they had specific duties that required them to be there or above. Also, if you did have to be on the hangar bay or above, you were supposed to wear your dress whites. Douglas remembers being on the 05 level (in the island) in his dungarees watching everything and nobody asked him why he was there or why he was not in his whites. Douglas also remembers bow the crew hazed some of the journalists and made them get navy haircuts—ruining their perfect television hair!

President Nixon left after greeting the astronauts and the *Hornet* headed back to Pearl Harbor. On July 26, 1969, the *Hornet* docked at Pearl Harbor and offloaded the astronauts in their MQF as well as all the other specialists, guests, and special equipment. The MQF (with the astronauts inside) was flown to Houston where the astronauts moved into a larger, more comfortable quarantine facility. The *Hornet* then departed for Long Beach once again.

Right: Apollo 11 command module being lifted onto elevator three by USS *Hornet*'s crane, July 24, 1969.

Below: Apollo 11 Command module on the deck of USS *Hornet* (CVS-12), July 24, 1969.

Above: Astronauts Armstrong, Collins, and Aldrin (left to right) congratulated by President Nixon through the window of the MQF in the hangar bay of USS *Hornet* (CVS-12), July 24, 1969.

Left: President Nixon talking to the Apollo 11 astronauts through the window of the MQF on board USS *Hornet* (CVS-12), July 24, 1969.

Above: Marine Joe Holt and the Apollo 11 command module on USS *Hornet* (CVS-12).

Right: USS *Hornet* (CVS-12) enters Pearl Harbor with the Apollo 11 astronauts, supporting technicians, and press. Many of the crew are lining the rails in dress whites, July 26, 1969.

The MQF is off-loaded from USS *Hornet* (CVS-12) with the astronauts still inside at Pearl Harbor, July 26, 1969.

The governor of Hawaii, John Burns, greets the Apollo 11 astronauts while they are still in their MQF at Pearl Harbor, in front of a large audience of press and dignitaries. USS *Hornet* (CVS-12) is in the background and banners are hung saying "Hornet +3" and "Apollo 11," July 26, 1969.

Apollo 12

From late August into mid-September, the *Hornet* stayed on the west coast. The ship and crew participated in a large ASW exercise (HUKASWX-9-69) and then performed carrier qualifications. In September, it was announced that the *Hornet* would be the recovery ship for Apollo 12. Considering how good of a job the crew had done with Apollo 11, this was not a big surprise. On October 27, the *Hornet* left for Pearl Harbor with most of the same equipment and specialists as last time. The big difference for this recovery was going to be less press and no presidential visit.

Apollo 12 was the first operational mission. Technically Apollo 11 was the last test flight—proving that astronauts could land on the Moon and be recovered. The team of astronauts was all navy this time: Charles "Pete" Conrad, Alan Bean, and Richard Gordon. Conrad was a naval aviator and test pilot who set an eight-day endurance record as pilot of Gemini 5, commanded in Gemini 11, and finally walked on the Moon during Apollo 12. Bean was also a naval aviator and test pilot, and his first space mission was Apollo 12 where he also became the fourth man to walk on the Moon. Gordon was also a navy aviator and test pilot who had been the pilot of Gemini 11 with Conrad and the command module pilot on Apollo 12.

While crews prepare the Apollo 12 command module for recovery, a SH-3D Sea King helicopter of HS-4 is lifting one of the astronauts from the raft on their way to USS *Hornet* (CVS-12), November 24, 1969.

After an eventful take-off that included two lightning strikes, the Apollo 12 rocket and crew made it to the Moon, landed, lifted-off, and returned. When they splashed down on November 24, 1969, the *Hornet* had managed to steam into an almost perfect position—within 2.5 miles. The recovery of the astronauts and capsule went just as smoothly as it had the first time and the *Hornet* signs this time read "Three More Like Before." The only mishap was a hard landing on the ocean that dislodged some equipment, causing a minor injury to Bean. There was much less fanfare than the first mission, but in many ways Apollo 12 was more impressive. The landing on the Moon and the splashdown back on Earth were much more precise. The mission demonstrated that Apollo 11 was not a fluke. The astronauts and all the special equipment were dropped off back at Pearl Harbor again and the *Hornet* headed home.

By December 4, 1969, the *Hornet* was back in Long Beach again. At the beginning of 1970, the *Hornet* was selected as the recovery ship for Apollo 13. However, Apollo 13 was delayed, and it turned out that Apollo 12 would be the last mission for the *Hornet*. On January 15, 1969, it was announced that the *Hornet* would be sent to Bremerton for decommissioning. This meant the *Hornet* would miss all the drama of Apollo 13.

This brought to an end a service record spanning from late 1943 through 1970, including two major wars, the Cold War, and three space capsule recovery missions. Few ships can claim an operational record as impressive as the *Hornet*, although some of her sister ships can probably make the best case. When the *Hornet* was launched, the navy had only recently transitioned from biplanes to monoplanes. She carried the pinnacle of piston-engined aircraft during the largest war in human history. The *Hornet* then added early jets and helicopters to her air groups, serving with distinction in the early part of the post-Korean War and Cold War. The *Hornet* then transitioned into an ASW platform, handling the most sophisticated anti-submarine aircraft of the time. Finally, the *Hornet* was transformed into a NASA recovery ship, bringing home the first and second groups of astronauts that walked on the Moon.

In this final phase of the *Hornet*'s active duty service, the ship was not modified for her new mission of picking up astronauts. However, she showed (as did the other *Essex*-class carriers that participated in the various space programs) that the design was one that allowed for significant flexibility. Without having to modify the ship, it was transformed into a space capsule recovery ship. The hangar bay space, communication equipment, radar, experience in working with helicopters, EAW aircraft, and more all made the *Essex* class a perfect fit for the space programs of the 1960s.

Right: USS *Hornet* (CVS-12) maneuvers into position to recover the Apollo 12 command module, November 24, 1969.

Below: Capt. Seiberlich congratulates Apollo 12 astronauts in the MQF on the hangar bay of USS *Hornet* (CVS-12), November 24, 1969.

6

Final Transformation into a Museum Ship

Decommissioning

In October 1969, the navy conducted an extensive inspection and survey (INSURV) of the *Hornet*. This led to the January 15, 1970 decision to decommission the *Hornet* as of June 30, 1970. As the ship prepared for decommissioning, the final arrested gear landing was made on February 29, 1970 by Commander Gerald Canaan and Rear Admiral Norman Gillette. Gillette graduated from the Naval Academy in 1936 and started World War II serving on the USS *Texas* (BB-35). He became an aviator in 1943 and then commanded a squadron of PBY Catalinas in the southwest Pacific for the rest of the war. Later on he was the commander of the USS *Thetis Bay* (LPH-6).

On March 2, 1970, the *Hornet* was at Long Beach where her deactivation was begun. All non-essential systems were removed, and once this was complete, a skeleton crew steamed the *Hornet* to Bremerton under her own power on March 30. On April 2, the *Hornet* arrived at Puget Sound Naval Shipyard and was officially placed "In Commission, In Reserve" status. The hull was painted, essential repairs were conducted, and all fuel and water was removed from the ship among other preparations for storage.

On June 30, 1970, the crew assembled for the decommissioning ceremony. The final captain of the ship, Seiberlich, was the last person to leave the ship. Once the ceremony was over, the real work of storing the *Hornet* could begin. This included cleaning all of the equipment and then making sure it was lubricated and covered in preservative. Elevator three was removed and stored on the flight deck. All openings and doors were closed and sealed, the guns were covered and the whole ship was humidified. The *Hornet* was then towed over to the Naval Inactive Ships Maintenance Facility (NISMF) where she would spend the next fourteen years along with other ships such as the USS *Bon Homme Richard* (CVS-31), USS *Bennington* (CVS-20), USS *Oriskany* (CVA-34), USS *Missouri* (BB-63), USS *New Jersey* (BB-62), and other smaller ships. Every ninety days, the ship was inspected. In 1973, 3 inches of protective foam was sprayed over the flight deck.

Officers and enlisted men leaving USS *Hornet* (CVS-12) after decommissioning in Puget Sound Naval Shipyard, June 26, 1970.

Capt. Carl J. Seiberlich is the final man to leave USS *Hornet* (CVS-12) and he gives a final salute, June 26, 1970.

In 1984, the *Hornet* was moved to a new spot in the same facility where she stayed for another ten years. In 1987, a full inspection for usability determined that all of the *Hornet*'s systems were obsolete. In 1988, the *Hornet*, along with some of her sister ships, was recommended for disposal, and on August 19, 1989, the *Hornet* was stricken from the navy's rolls.

Just before the *Hornet* was stricken from the navy's rolls, a group called the "USS *Hornet* Historical Museum Association Inc.," (formed in March 1989) applied to have the navy add the *Hornet* to the navy's ship donation program. In December 1991, this association helped get the *Hornet* designated a National Historical Landmark. However, they could not raise enough funds, and in January 1993, the *Hornet* was sold for scrap to Astoria Metals. As we will see later, this designation of the *Hornet* as a Historical Landmark was going to be key in saving her from the scrapyard. The ship was towed to Hunter's Point in San Francisco to begin the scrapping process in October 1994. Astoria Metals struggled to put together a plan to scrap the ship under budget and within the time limits set by the navy.

The official designation of the *Hornet* as a National Historic Landmark reads:

> In 1991 the *Hornet* was designated a National Historic Landmark both for its service in the Pacific in World War II and as the recovery ship for the Apollo 11 and Apollo 12 astronauts. Involved in several of the heaviest and most critical battles in the Pacific, the *Hornet* earned seven battle stars, and a Presidential Unit Citation. The *Hornet* or its aircrews were responsible for destroying more than 1400 enemy aircraft. After

Adml. William F. Bringle, commander naval air force, US Pacific Fleet, talks to Capt. Carl J. Seiberlich during the decommissioning ceremony of USS *Hornet* (CVS-12), June 26, 1970.

> undergoing modernization the ship was called to service in recovering the landing capsules for the Apollo space program. The Apollo 11 mission was the first landing on the moon. The *Hornet*, with President Nixon on board, picked up astronauts Neil Armstrong, Edwin "Buzz" Aldrin, and Michael Collins from the sea in the summer of 1969, welcoming them to earth from their historic mission. The *Hornet* recovered the astronauts of Apollo 12 before the carrier was de-commissioned in 1970.

The registration form was filled out and submitted by James P. Delgado, executive director of the Vancouver Maritime Museum.

While this was going on, Alameda Naval Air Station was in preparation to be closed down and transferred to the city of Alameda in 1994. Navy Captain Jim Dodge, commander of the base, wanted to have a months-long closing ceremony culminating during Fleet Week. He had the perfect idea for a ship to use in this: the *Hornet*, just across the bay and not yet scrapped. He asked Astoria Metals if he could "borrow" the *Hornet* for this ceremony and they happily agreed since this extended their deadline to figure out a workable plan to scrap the ship. There was some legal wrangling with Navy Sea Systems Command (NAVSEA) and the Defense Revitalization and Marketing Service before this was approved but eventually everything was sorted out.

On May 11, 1995, the *Hornet* was towed to Alameda Naval Air Station and berthed next to USS *Carl Vinson* CVN-70. The *Hornet* looked very different from her larger, cleaner cousin. This was the first visit back to Alameda for the *Hornet* in almost thirty years. Large numbers of volunteers came aboard and began to clean the ship in preparation for the open house. There was years of accumulated debris that had to be cleared off. The flight deck and island were made presentable in time for the start of the celebration of the closing of the station that started on May 20, 1995.

The *Hornet* opened for visitors and over 9,000 guests came aboard. Only the flight deck and island were available, but people were happy just to see and experience that. By July, the hangar bay and parts of the second deck had been cleaned and opened as well. The enthusiasm of the guests helped to spur a second attempt at saving the *Hornet*. A new organization, the Aircraft Carrier Hornet Foundation (ACHF), was created. As they began working on their plan to rescue the ship, the celebrations on board continued. The fiftieth anniversary of Victory over Japan Day was celebrated and later Fleet Week. During Fleet Week, the ship was open for four days and the *Hornet* had over 30,000 visitors! Events included honoring of three surviving Doolittle Raiders and 100 survivors of other ships involved in the raid. During these events, the ACHF was selling T-shirts and collecting signatures of people pledging to support saving the ship.

After Fleet Week was over, Astoria Metals asked NAVSEA to consider giving ACHF more time to raise the funds to save *Hornet*. By this time, Astoria realized there was no way they could scrap the *Hornet* and make any profit, primarily because of tough, California environmental laws. NAVSEA said no. However, because the *Hornet* had been designated as a National Historical Landmark, ACHF sued NAVSEA, claiming they needed to complete a Section 106 inspection before they could decide what to do with the *Hornet*.

According to the US General Services Administration's own website:

> Section 106 of the NHPA requires that each federal agency identify and assess the effects its actions may have on historic buildings. Under Section 106, each federal agency must consider public views and concerns about historic preservation issues when making final project decisions.

This was part of the National Historical Preservation Act (NHPA) of 1966. In October 1995, NAVSEA agreed to put *Hornet* in the donation program again for one year.

In November 1996, ACHF delivered their proposal for saving the *Hornet*. No other organization applied and NAVSEA accepted the plan and placed the *Hornet* in a special donation hold status. Their own evaluation of Section 106 indicated that the *Hornet* should have never been sold to Astoria Metals and that sale was nullified. Over the next year, the ACHF raised enough money to pay back Astoria Metals (plus interest) and to complete the work needed to open the *Hornet* officially as a museum.

Above: USS *Hornet* flight deck before restoration, 1995.

Right: USS *Hornet* flight deck after restoration, 1995.

USS *Hornet* hangar bay before restoration, 1995.

USS *Hornet* hangar bay after restoration.

USS *Hornet* event, July 1996.

USS *Hornet* fundraising, July 1996.

Finally, on October 16, 1998, a huge gala was held on board the *Hornet* and thousands of people danced to live music in preparation for opening day the next day. On October 17, 1998, the *Hornet* completed her final transition into a museum as she was opened to the public. On that day, a large recommissioning ceremony was held that thousands of people attended. This ceremony included flyovers from the History of Naval Aviation Show and the Blue Angels.

During this transition, the *Hornet* changed but in a very different way from before. The ship had been demilitarized and preserved for a long time. This meant the amount of clean up and restoration needed was immense. Most of the major systems no longer functioned (the engines for example) but most of the lesser ones did after some work (such as the electrical system). However, the ship was not generating her own electricity anymore nor was she desalinizing water so she had to be connected into the regular water and electrical grid of the city of Alameda. The *Hornet* was ready for her final mission as an educational facility instead of as a warship. Fifty-five years after launching, she had found a new way to serve the community.

7

USS *Hornet* Museum

The USS *Hornet* Museum has been open for over twenty years now. During this time, it has undergone continuous restoration while simultaneously adding to the impressive list of artifacts in its collection. A large crew of volunteers and paid staff has dedicated themselves to preserving the history of the ship and each year new portions of the ship are opened for the public. On an average year, over 20,000 students come through the ship on field trips to learn about history and STEM. The ship hosts community events such as the anniversaries of the Apollo missions, the Doolittle Raid, the Battle of Midway, and holidays such as the 4th of July, Veterans' Day, and Memorial Day. The ship also serves as a large venue for private events such as corporate meetings, holiday parties, and many others. Finally, due to the unique history of the ship, the museum is a resource for veterans who volunteer in large numbers to work in restoration, security, education, and as docents. They enjoy being on the *Hornet* for many reasons but also because they can spend time with other veterans like themselves who understand their experiences.

Location

The ship is located at historic Pier 3 in the former Alameda Naval Air Station. It is from this same set of piers that USS *Hornet* (CV-8) left for the Doolittle mission in April 1942 (*CV-8* actually left from Pier 2). A large plaque at the entrance to the pier commemorates this event. A huge anchor of the same type that the *Hornet* used is also at the entrance. The airfield, hangars, engine testing facilities, and more of the former base can still be seen from here. Many of these buildings have been converted to civilian businesses now. The San Francisco skyline can be seen across the bay on most days, as well as the San Francisco/Oakland Bay Bridge and Treasure Island. The Oakland skyline and the Berkeley hills are visible as well.

USS *Hornet* Museum at Pier 3 Alameda, Ca.

A walkway down the side of the ship leads to the guest entrance, and from here, the large size of the ship becomes evident. The almost 900-foot length of the ship seems to go forever. The island raises high above from the top of the flight deck and the radar dishes on top make it even taller. On the side of the island is the very large number "12." Here and there are water marks and rust showing against the grey paint. A ship this size requires almost constant maintenance to keep all of the large ship in top shape. There are three brows (ramps) to the ship. The first is the entrance, the second is the exit, and the third is part of the loading dock. From the outside you can see the replenishment stations, elevator three, the crane, the starboard gun mounts, and more. Even from the pier, there is a lot to see.

Entrance/Hangar Bay One

The steps to the first ramp ascend to the officer's quarterdeck. When the ship was active, this was only used by officers and special guests. The entrance opens up to the admission desk in hangar bay one. There is a large shield here that depicts the uniform patch that was worn when the *Hornet* was designated as a CVS. The original ship's bell is located here (formerly mounted in the Fo'c'sle). To the right are the stairs that lead up to the captain's in-port quarters that are on the gallery or 02 level (under the flight deck but above the hangar bay). The captain would use these quarters when in port and there was an attached galley, head, conference

USS *Hornet* Museum entrance.

room, and office. This cabin could also be turned over to visiting VIPs. When in port, the captain had a room in the island on the navigation bridge. Right next to the stairs is the 1-MC room (for ship-wide announcements). The admissions desk is here and this is where the self-guided tour maps are given out. To the left is a raised projection room that was used to show movies on a large screen that could be lowered in the hangar bay. There are four small, closed openings where the movies would be projected through.

Past this is the main part of hangar bay one which is the hangar bay closest to the bow of the ship. The ceiling is high above and the entire area is much larger than it looks from the outside. Overall the entire hangar bay is 654 feet long, 70 feet wide, and anywhere from 17 feet 5 inches high (where the gallery deck is) to 27 feet high (where it goes all the way to the flight deck).

The hangar bay has many tie downs sunken into the floor to hold aircraft in place although most of them have been filled in or covered over for the safety of guests. In all three hangar bays, there are hose reels, nozzles, and piping in three different colors: purple, red, and green. Purple is for aviation fuel, red is for firefighting (seawater), and green is also for firefighting but with fog/foam.

To the immediate right is the *Hornet* battle board from World War II. It lists all the battles participated in, all the planes shot down, and all the ships sunk by the *Hornet*. This is a recreation of the original that resides in a different museum.

Left: Movie projection room and ship's bell in hangar bay one.

Below: Captain's in-port cabin, up stairway from hangar bay one.

The largest bomb elevator (big enough to handle nuclear weapons) that goes from the third deck ready magazine to the hangar bay and flight deck is visible here as well. Beyond the battle board are several aircraft (covered elsewhere in this book).

Across the hangar bay is the orientation area. The orientation area holds chairs and a large monitor. A short video explains the history and rules of the museum and then the docents on hand can talk about available tours and answer any questions guests might have. There is a large map of the Pacific War and a small photo exhibit behind the orientation area. This is the "Hangar Bay Life" exhibit that features photos of from ordinary sailor life in the hangar bay from 1943–1970. There are several excellent pictures in there including men playing basketball in the elevator shaft and the thousands of cots in the hangar bay during Operation Magic Carpet.

A large, somewhat worn flag that flew from the *Hornet* on February 16, 1945 is hanging on the wall in a glass case on the starboard side. Guests can head down to the second deck and the exhibits there or move forward to the Fo'c'sle from hangar bay one. Large side doors can (and often are) opened to the outside to let fresh air and sunlight in. A feature that can be seen here are that the ladders going down have a "coaming" around them which is a raised 9–12-inch metal ridge to prevent spilled fuel, water and other materials from going down to the lower decks. These coamings are on all the ladders going down from the hangar bays.

The most impressive feature in hangar bay one is the large elevator shaft that is for elevator one. This elevator is almost always in the up position showing how big the elevator shaft is. Inside the elevator shaft are the workings of the large elevator. This includes the CONFLAG station which is a fireproof and blast-proof room with viewing slits and communications equipment from which firefighting could be directed. There is one in each hangar bay. There is a water curtain here that would release a solid sheet of water at the entrance to elevator one to try to prevent a fire in the hangar bay from entering the elevator well or the reverse. Several of the cables that are used to raise and lower the elevator as well as small orange rods that are automatically raised around the elevator opening on the flight deck when the elevator is down are visible here. This elevator still works and is still used to move aircraft and other heavy equipment up and down from the flight deck.

There is also a ladder on the port side that leads to the admiral's in-port cabin—across from the captain's quarters on the gallery, 02 level. Similar to the captain, the admiral had another cabin in the island for use when at sea. The admiral also had his own head, galley, office, and conference room similar to the captain. In addition there was a small medical room for use by both the captain and the admiral if needed. On this same side there is a large shield that shows the patch worn by the crew when the *Hornet* was designated as a CVA. Near this is one of the fog/foam stations. On the deck next to this is an ammunition hoist for the 5-inch guns on the port side. This is an easy object to miss as it looks like an innocuous protrusion of some type from the hangar bay deck. The large pulley above this was used for raising pallets of ammunition.

Admiral's in-port cabin, up stairway from hangar bay one.

Admiral's in-port conference room, up stairway from hangar bay one.

There is an exhibit of a pilot's flight suit (*circa* 1960s–1970s) in a glass case that can be moved but is usually in hangar by one. This flight suit was donated by friend of the ship and naval aviator Willie Sharp. Willie Sharp is a frequent guest on the *Hornet* and he has quite an amazing story of avoiding capture off the coast of Vietnam. This story is well documented in other books.

Aft from hangar bay one is hanger bay two. There are massive, open doors between the two hangar bays that can be closed to separate them in case of fires and other issues and there are also two more water curtains.

Hangar Bay Two

In hangar bay two, the ship's store is on the starboard side. The store offers books, apparel, and many other museum-related items. Behind the store (now used as an office) are the old photo labs. Just beyond the store is the exit and security station. There is also a ladder to the flight deck here. Right next to this is CONFLAG station for hangar bay two. Adjacent to this is the slanted bulkhead that carries the exhaust from the fire rooms away.

There are more aircraft here as well as elevator two and the hangar bay control room. If the doors are open here, elevator two can be seen on the port side. This elevator is also typically in the up position so normally just another ocean view will be visible and the underside of elevator two. Elevator two also still works and is still used to transport aircraft and other heavy objects back and forth from the hangar bay and flight deck. Inside the hangar bay control room, located so it can look over the hangar bay, is a scale model of the entire hangar bay and small aircraft models. From here the movement of all aircraft in the hangar bay could be monitored and directed.

Further aft, there is an opening that leads to the escalator added for pilots during the refit in 1956. This is the easier way to the flight deck. More firefighting stations are located here as well. To the left and below the funnel is a space used for rotating special exhibits. These typically last three to six months and are often associated with special ship events.

The main attraction in this hangar bay is the Apollo exhibit, and there are multiple artifacts here, including the MQF from Apollo 14, CM-011 (the block one command module that the *Hornet* picked up in 1966), and a SeaKing helicopter that was actually used as a movie prop for Apollo 13. On the floor are painted footsteps that were the first footsteps on "land" for Neil Armstrong after splashdown and the helicopter ride to the *Hornet*. There are banners hanging from the ceiling here with large pictures from the Apollo missions, including the famous one of President Nixon talking to the astronauts through the window of the MQF. A BIG suit display has recently been added.

In a room off the starboard side of hangar bay two the Apollo exhibit continues. This room has numerous artifacts from the Apollo missions donated to and collected by the museum, including a vintage TV set showing footage

CONFLAG station in hangar bay two.

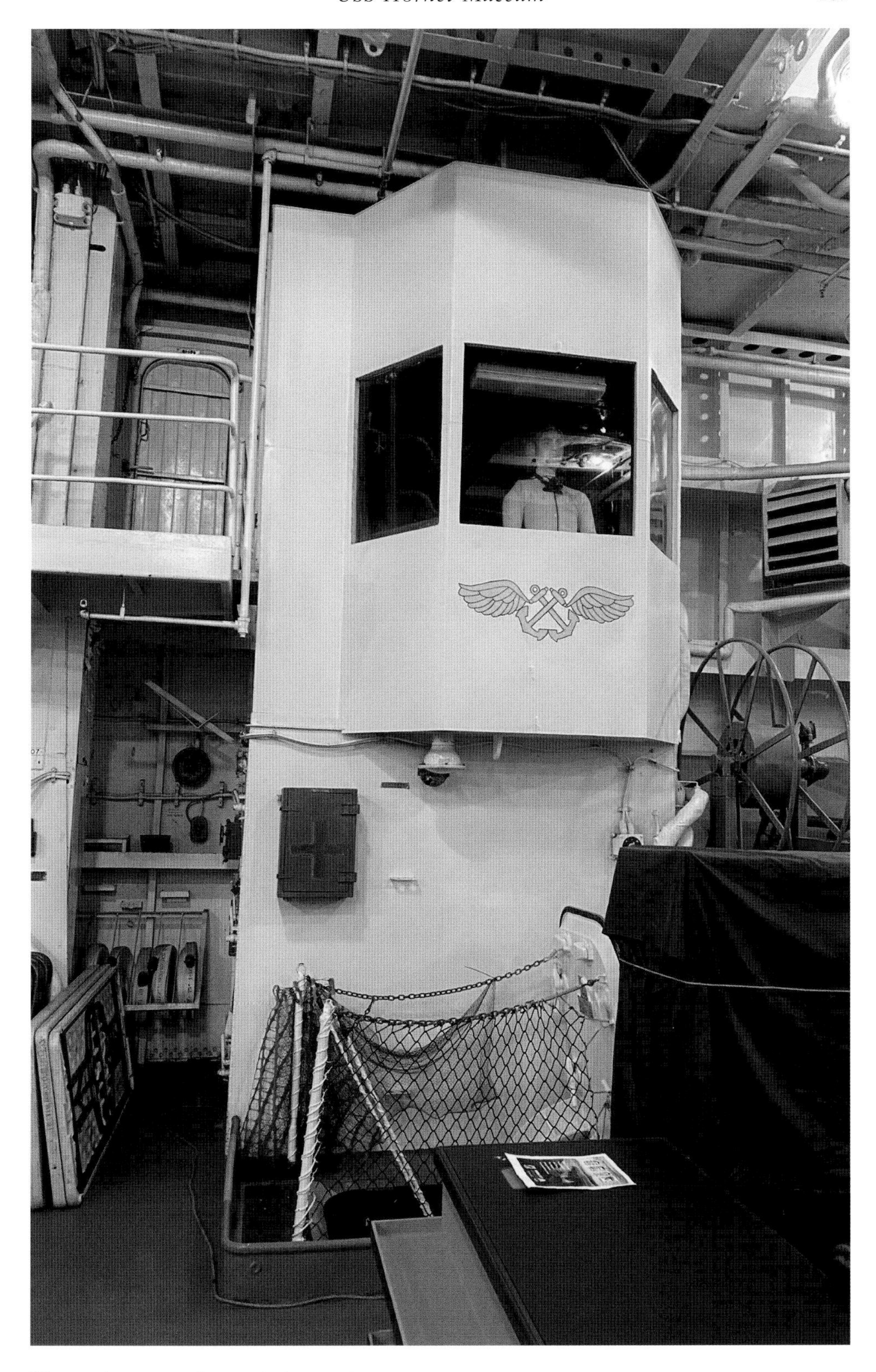

Hangar bay control room in hangar bay two.

Aviation fuel station in hangar bay two.

from Apollo 11 and some scenes from the *Hornet*. Behind CM-011 there are other, smaller information displays about the Apollo programs. These include a flight jacket from one of the members of the helicopter squadron that recovered the astronauts (HS-4 Black Knights), an Apollo couch (where the astronauts sat while in the capsule) from Apollo 10, and a few displays about the design and construction of the re-entry vehicles.

There is a bomb elevator right outside the entrance to the Apollo exhibit that still works and is used by the museum staff to bring moderate sized equipment to the flight deck or down to the third below. The large heavy door to the bomb elevator is typically closed but guests can see the handle and cable used to open it.

Hangar Bay Three

Proceeding into hangar bay three through the large, typically open doors there (with two more water curtains), there are more aircraft as usual. This is the largest hangar bay of the three. On the port side is a large mural that was painted to show most of the important aircraft that served on the *Hornet*. Next to this is an opening that leads to the ship's theatre that was used primarily for training films when the ship was active. This theatre is being refurbished with the idea to use it to show short films to public while visiting the ship.

On the starboard side is a large door that is almost always open showing elevator three, in the down position. That is used as a loading dock for the ship and right outside is the large crane that lifted the Apollo 11 and 12 command modules onto the ship and is still used to lift heavy objects onto the ship such as aircraft. This elevator no longer functions, so is always in the down position. A powered lift next to a ramp lets this area be used as a handicapped entrance/exit and as a way to load smaller items on the ship.

Further aft in hanger bay three is a large flight simulator on the port side and then a large open area near the restrooms. This is the main event area and typically has a stage set up for whatever event is coming up next. The stage has a large American flag hanging behind it and can be connected to a sound board and speakers for presenters to use. Many configurations of tables and chairs can and often are set up here for events. This area is also frequently used for temporary exhibits. If guests look up in this area they will see racks holding external aircraft fuel tanks and other equipment.

On the starboard side are large shields, beautifully painted, holding the insignias of many of the *Hornet*'s sister ships. Included are shields for USS *Essex* (CV-9), USS *Intrepid* (CV-11), USS *Randolph* (CV-15), USS *Lexington* (CV-16), USS *Bunker Hill* (CV-17), USS *Bennington* (CV-20), USS *Shangri-La* (CVA-38), USS *Hancock* (CV-19), USS *Bon Homme Richard* (CVA-31), USS *Oriskany* (CV-34), USS *Philippine Sea* (CV-47), USS *Wasp* (CV-18), USS *Kearsarge* (CVS-33), USS *Lake Champlain* (CV-39), and USS *Ticonderoga* (CV-14).

Near the back is the aircraft restoration area. There is almost always at least one aircraft here being restored. This area is sectioned off but on many days the aircraft restoration crew can be seen working on the aircraft. Their workshops are behind this area. Above this is the CONFLAG station for hangar bay three.

A few rooms off the starboard side of the hangar bay have been restored. These include the tire repair ship and the aviation electrical shop. Some interesting placards can be seen here on the outside of rooms such as the "R.S. Bomb Fuse Locker." Various firefighting stations can be seen in here as well. Elevator three used to be in the middle of the hangar bay before it was moved to the side during one of the refits. This created lots of issues with storing and moving aircraft and the area really opened up once it was moved. There is a large tackle and pulley system on the ceiling here—to lift heavy objects such as engines.

A ladder and catwalk at the back wall of hangar bay three leads to the multi-room "Nisei Veterans' Exhibit." This includes the 100th Battallion/442nd Regimental Combat Team and the Military Intelligence Service and is hosted by the Friends and Families of Nisei Veterans. This large exhibit has a wealth of information about Japanese-Americans in World War II. It includes a small theatre-style room where videos are shown as well as a huge collection of artifacts. The full military history of Japanese-Americans is presented as well as a lot of information about famous veterans of these units who have gone onto long careers in the military and in politics. On most weekend days, there are docents

Aircraft restoration in hangar bay three.

here who can provide extra information about the exhibit. There is so much information here that guests could take a full day just to visit this portion of the museum. This museum within a museum is a result of a long partnership between the Friends and Families of Nisei Veterans and the USS *Hornet* Museum.

All the way aft is the fantail. This area has been beautifully restored and has one of the best views of the San Francisco skyline anywhere. Many memorial events are held here that culminate in a wreath toss into the ocean. As the ship's chaplain, Whatley, puts it: "This is the most important station on the ship. Whoever is on lookout here is the last chance for anyone who has fallen overboard to be seen and rescued."

Above is the safety net that would catch anyone blown off the aft portion of the flight deck. A large reel and tripod here holds a "Fanfare" unit. This was a sound and echo acoustical decoy that could be deployed up to 600 feet away from the ship in the water to try to trick submarines and torpedoes. This was added when the ship was an ASW carrier. There is also a large, metal bell, life-preserver rings, and large loops of rope.

Above: Friends and Family of Nisei Veterans exhibit in hangar bay three.

Right: Fantail off of the hangar bay.

01 Level

The 01 level is forward of hangar bay one and can be accessed from there or from the second deck. From hangar bay one, there are entrances on both sides of the elevator well. Straight ahead are various officers' berthing areas. A ladder ascending one of two flights of stairs leads to the junior officer's berthing (JOB) area. There is a head here that services this whole area.

Forecastle (Fo'c'sle)

The main open area that houses the anchor chains is just ahead. This is the Fo'c'sle. The anchors themselves weigh 15 tons. Each anchor chain is well over 1,000 feet long and is made up of 120-pound links. One end of the chain is attached to a heavy duty chain locker below and the other end connects to the anchor. The chains wrap around a large brass winch and there are two smaller hand wheels next to the winch. These are called "Wildcat Brake" and "Control Handwheel." These are used during the raising and lowering of the anchors. The ship's bell located at the entrance used to hang in this area. There are pictures in here that show the bow as it looked after it suffered damage from the typhoon in World War II.

There are two rooms above this area that can be reached from here. The first one is one of the largest berthing areas on the ship. There are 257 bunks here, three or four high. There is only one large head to support all of the sailors berthing here (mostly catapult operators and plane handlers). Each sailor had a small locker for their personal effects that was only 24 inches by 24 inches by 12 inches in size.

The second area that can be reached is from the forward "Hurricane Bow" (so named after it was enclosed) that has the bullnose (very front of ship), seven portholes, a 5-inch gun replica for training, and then the ladder to the secondary conn. This are is not open to the public but can be seen. During battle (or general stations), the executive officer (XO) would be stationed here. If the bridge was damaged or the captain injured, the XO could command the ship from here. It has all the same equipment as the bridge making it fully redundant. That includes a radar repeater, wheel, and the engine order telegraph.

Officer's Barbershop

Forward on this level is an area only accessible on a tour with a docent. Besides more officer's berths there is a small officer's barbershop complete with two barber chairs, mirrors, some vintage magazines and a sink.

Officer's barbershop on the 01 level.

Sonar Room

Beyond the barbershop, there is another room that houses the sonar room that was added in when the *Hornet* was converted to an ASW carrier. Some of the original sonar equipment is still in here. Because of its stability, the carrier could be a more effective sonar platform than a smaller ship like a destroyer. However, carriers are usually moving fast to operate aircraft and that diminished the effectiveness of the sonar equipment.

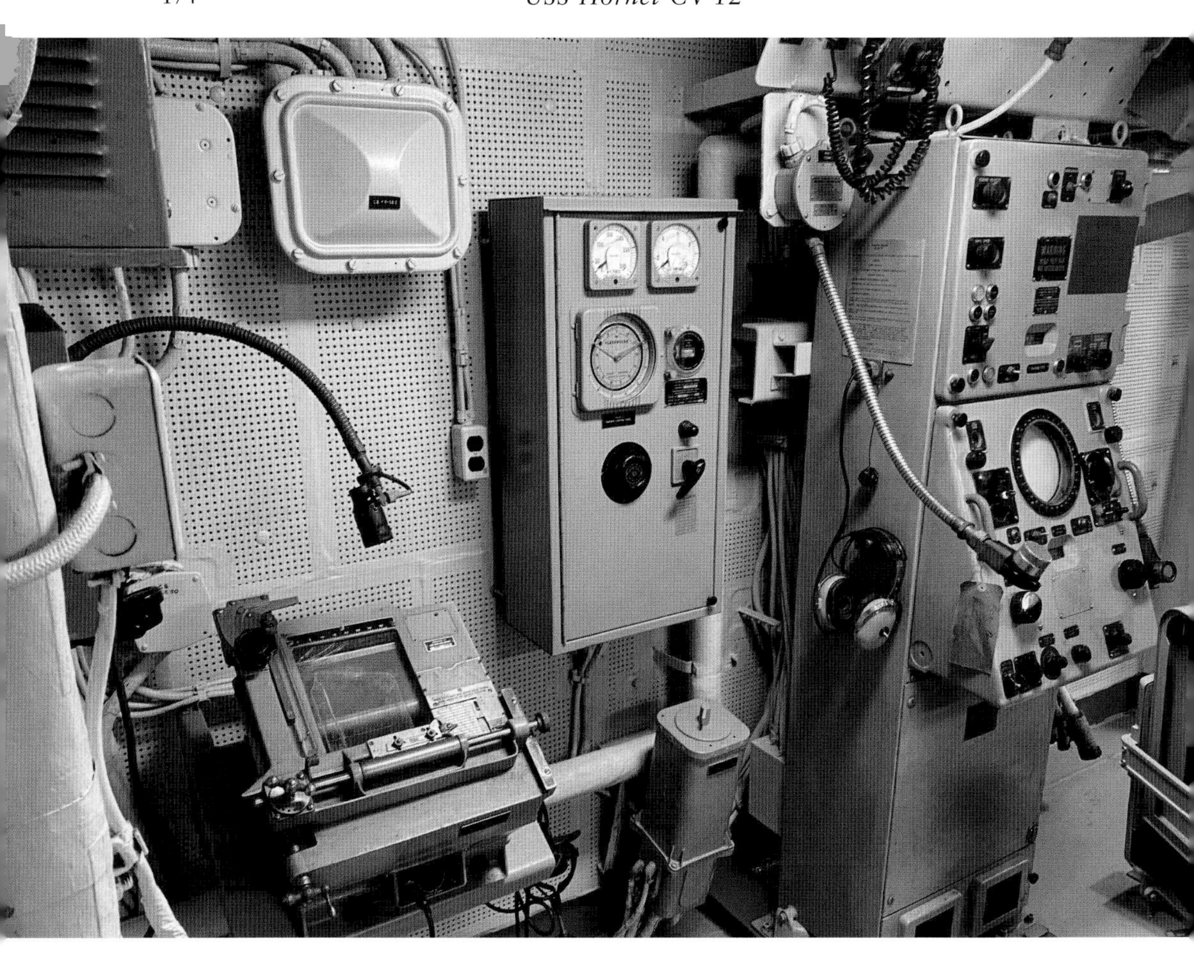

Sonar room on the 01 level.

Gun Mount

There is a ladder from the Fo'c'sle to the forward, starboard restored gun mount that has a fully restored Mk. 24 5-inch/38-caliber and twin Mk. 33 3-inch/50-caliber guns. There is also the Gun Director Mk. 56 next to the guns (added after World War II). By tying in with the ship's radar in the radar room complex (located inside behind the guns), the guns could be trained to fire at target with much more accuracy by gauging their height, direction, speed and other factors such as wind. Combine this with the top secret (in World War II) proximity fused shells, and US Navy ships could put up a strong anti-aircraft defense. Right behind the guns is the ammunition ready service room. This has four racks for the five inch guns that would be kept ready during combat. Ammunition could be brought up from below by using two ammunition hoists. These guns were not on the *Hornet* when she was donated to the ACHF and they had to be found and replaced from other ships in the San Francisco Mothball fleet on Suisun Bay.

Forward, starboard 3-inch anti-aircraft guns below the flight deck.

Forward, starboard 5-inch anti-aircraft gun below the flight deck.

Forward, starboard Gun Director Mk. 56 below the flight deck.

Flight Deck

From hanger bay two, the flight deck can be accessed via the ladder next to security or the escalator. On the flight deck there are typically several aircraft at various places. The flight deck itself is somewhat recognizable from the movie *XXX: State of the Union* that starred Ice Cube. To the right of the island and towards the bow of the ship are two bomb elevators. The larger one is visible below in hangar bay one and is big enough for nuclear weapons. Along the starboard side are many poles that are off the side of the ship—some are vertical and some horizontal. These are radio antennas and they had to be movable so aircraft could take off without hitting them. Some of them still work and are used by the *Hornet* Amateur Radio club (detailed later in this book).

The gun mount accessed from the Fo'c'sle is visible from above. Elevator one is visible from here-a large square right in the center of the flight deck. When in the up position it is actually part of the flight deck. From here to the front of the ship are the two catapults used for launching aircraft. Also on the starboard side is the catapult deck edge control panel. From here the operator could add tension to the bridal (attached to the aircraft), push the fire button to launch the aircraft and retract the shuttle after aircraft launch.

Walking across and then down the other side is the port, forward 5-inch gun mount. Behind the catapult on the port side the blast deflector is usually in the up position here. This is the position it would be in when aircraft were about to launch-to protect deck crewman from the back blast of jet engines.

Further towards the stern of the ship is the port side elevator two and then the angled flight deck for landing aircraft can be seen. Elevator two is actually an extension of the landing portion of the flight deck when it is in the up position. A painted white line down this section shows the landing aircraft where the centerline of their landing strip is. There are also helicopter landing circles scattered around.

Continuing aft there are four arrestor gear cross deck pendant boxes. These held the arrestor wires the planes used when landing. Guests can see an example of an arresting wire and an aircraft tail hook on the side of the island to get an idea of the size and strength of both. Off the port side of the ship is the Fresnel lens optical landing system (the docents can even turn it on for guests so they can see the lights). From the mid-1950s onward, all navy pilots use some version of this system to assist them in landing. The restoration project to make these lights work was a large one and much of the equipment (such as wiring) had to be replaced to make it work.

Further aft is the stern port-side 5-inch gun mount and Gun Director Mk. 56. These are in need of restoration and will be done at a future date. Beyond this is the landing signal officer (LSO) platform off the side. This includes a net where that officer can jump if there is a bad landing. At the rear of the ship is a flagpole that is only used when the ship is in port (since it would interfere with aircraft landings otherwise). The port side of the safety net mechanism (used to try to trap

Catapult deck edge control panel on the flight deck.

Catapult blast shield on the flight deck.

Landing control officer station on the flight deck.

aircraft that might have arrestor hook issues during landing), can be seen here off the edge of the flight deck.

Going forward on the starboard side is the rear 5-inch gun mount (and associated Gun Director Mk. 56), and then elevator three. This gun has been recently restored and it looks brand new. As mentioned, elevator three is permanently in the down position and is used as a loading dock for the museum now. The starboard side of the safety net mechanism can be seen here and it is on the flight deck instead of off the side. In addition the large crane is clearly visible from here next to the elevator. Just beyond this is the smaller bomb elevator that was visible from hangar bay two below.

Up on the flight deck, high above the water, there is a great view of the San Francisco Bay Area. In addition to everything that can be seen from the fantail there is also a great view of the East Bay. The ship holds summertime events up on the flight deck as the large surface area is ideal for large crowds.

The flight deck.

Island

Now back to where the flight deck was accessed is the island. Tours of the island are docent-led only. The base of the island is 03 level (the level below the flight deck but above the hangar bay is the 02 level). Going up the ladders from here are multiple levels starting with one of mostly radar rooms and storerooms (04 level), then the flag bridge (05 level), the navigation bridge (06 level), the anti-aircraft control deck (07 level), and primary flight control (08 level). You can also go down from the island to the 02 level to see radio central, the combat information center (CIC), and other associated areas.

From the outside of the island, several important details can be seen. There is a display of the different color vests worn by the deck crewmembers as well as a large arrestor hook. Next to that is a small display explaining what each color means:

Purple was for those responsible for aviation fuel.
Blue was for plane handlers, aircraft operators, tractor drivers and messengers.
Green was for those responsible for crewing the catapults, maintenance of the aircraft, cargo-handling, ground support equipment, hook runners, photographer's mates, and helicopter landing signal personnel.
Yellow was for aircraft handling officers, catapult and arresting gear officers and plane directors.

Red was for ordnance men, crash and salvage crews, and responsible for explosive ordnance disposal.
Brown was for air wing captains and air wing line leading petty officers.
White was for air wing quality control personnel, squadron plane inspectors, landing signal officers, air transfer officers, liquid oxygen crews, safety observers, and medical personnel.

A large diagram showing the different configurations of the *Hornet*'s flight deck is also hung here. Higher up is a painted board with the campaign ribbons given to the *Hornet*, two capsules painted on the side representing the successful recovery of Apollo 11 and 12 and a small President's Seal showing where President Nixon stood to watch the Apollo 11 astronauts be brought on board.

Higher up the windows are visible that show where the flag bridge, navigation bridge, and primary flight control are. On top of all this are the large mast and the two radar dishes. The first dish is a large rectangular one and that is the AN/SPS-30 height finding radar and the other one that looks like a large satellite dish is the AN/SPS-43 air-search radar. Below and in front of these you can see the large ship's spotlight. Below and behind these is the funnel for exhaust from the

Island as seen from the flight deck.

Front view of island showing the windows from the navigation and flag bridges.

fire rooms below (also seen in hangar bay two), and below that a large radome with SPN-35 radar inside that was used to guide planes in to land. Right behind the radome and below primary flight control was the photo platform that was used to photograph all take-offs and landings. There is also the large twelve painted on this side of the island as well.

03 Level

Going inside, the first floor is the 03 level and this housed the flight deck crew shelter and flight deck control room. In the flight deck control room, there is a small exhibit describing how the movement of aircraft on the flight deck was communicated through and tracked here. This includes the "Ouija Board" map of aircraft on the flight deck, similar to the hangar bay control room below. These areas are currently used as break areas for the docents stationed on the flight deck as well as for storage. The room at the top of the escalator has some information regarding aircraft carriers and flight operations on the wall. From here are the ladders that let guests access all the other levels in the Island.

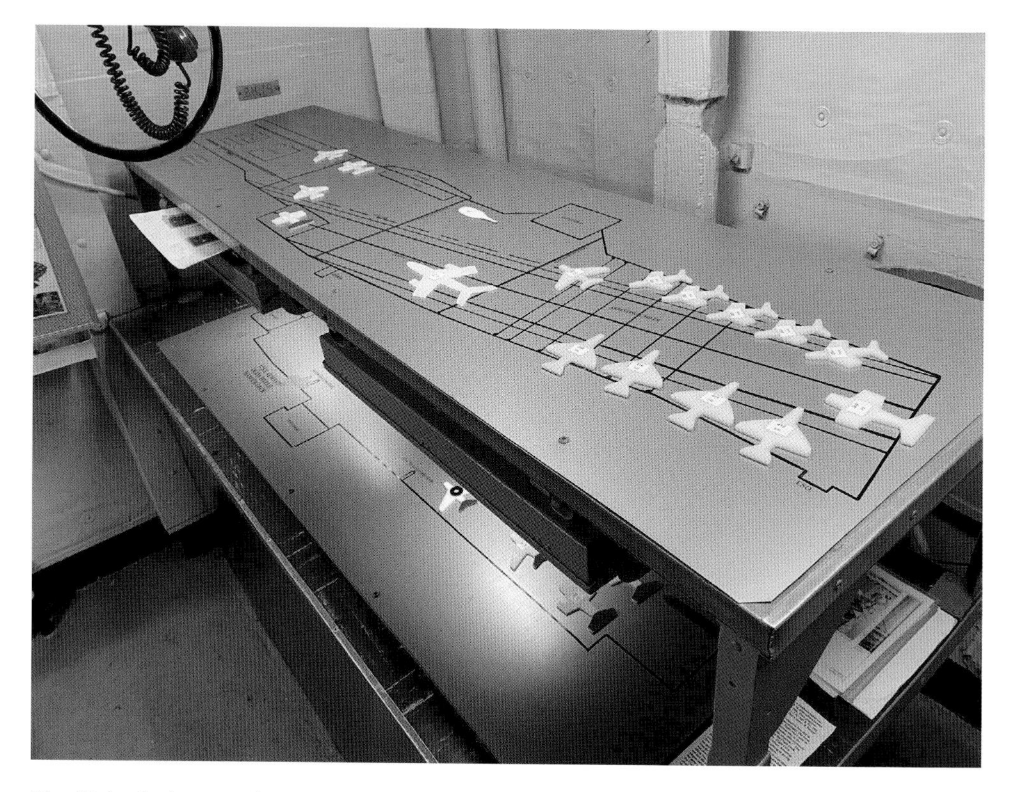

The flight deck control room Ouija board at 03 level of island.

04 Level

The 04 level is mostly store rooms and a radar room and not open for tours. On occasion, one of the doors will be open and some of the radar equipment is visible there. Specifically "Radar Equipment Room No. 2" is available to look into and it contains the transmitter/receiver for the SPS-30 height finder radar above the island. Otherwise it is just a stop on the way to the higher levels.

05 Level Flag Bridge

The next level is the flag bridge (05 level) that had an at sea cabin for the task force commander and was where he would operate with his staff. You can see his at-sea cabin here as well as some of the same control and navigation equipment available on the navigation bridge (described below). In addition there is some more space for the admiral to work with his staff as well as fleet deployment information. There is a framed picture of the *Hornet*'s most famous admiral, J. J. "Jocko" Smith, here.

06 Level Navigation Bridge

The next level up is the navigation bridge on the 06 level. Here the captain had an at-sea cabin and this is where he ran the ship from. On this level is the chart room (more specifically referred to as the "Captain's Plot & Message Center"), which housed the navigation officer and staff. They had a dead reckoning trace (mechanical tracking of ship's position), long-range radio navigation system (LORAN), a chronometer, and, yes, a sextant. It was their job to know exactly where the ship was at all times and to do this they used the mechanical equipment they had as well as every four-hour sextant readings to confirm and correct their position if needed.

After the chart room on the same level is the pilothouse. The equipment in here included the engine room indicator, ship's wheel, and the entrance to the captain's at-sea cabin. The ship's degaussing panel was also here (to demagnetize the ship against magnetic mines and torpedoes). Finally, the engine shaft tachometers are here, showing how fast the propellers are turning.

In front of the pilothouse is the bridge itself. On the starboard side sits an auxiliary conn and a seat with a good view to watch at-sea replenishment operations. Around the front of the bridge are all the communications needed to communicate anywhere on the ship including 1MC (full ship announcements) and 5MC (flight deck only announcements). Direct lines to the CIC and the aviation control tower are here as well. The captain's chair is on the port side of the bridge and only the captain sat here. Above this seat is a television camera where the captain could watch aircraft operations.

Also on this level is the flag signal platform. In certain situations, the ship still needed to retain the ability to communicate via signal flags or with signal lights using Morse code just like ships of the past. There is an identical platform on both sides of the island.

07 Level Anti-Aircraft Control Deck

The next level is the anti-aircraft control deck (07 Level). This level is only partially open via tours, but it housed the anti-aircraft control room and weather office, responsible for coordinating anti-aircraft fire as well as collecting and sharing all weather information for the ship.

The weather office is more correctly called the aerological office and lab and was where all the weather information was tracked and shared. In this room are charts showing weather patterns, cloud types and more. There is a small library of weather related books here. All of this was very critical information for ship and aircraft operations.

Navigation bridge chart room at 06 level of island.

Navigation bridge pilot house at 06 level of island.

08 Level Primary Flight Control

Primary flight control, or PriFly, is the next level (08). From here all aircraft operations could be directed. This is essentially the flight tower similar to any larger airport. The position of the windows here allow for viewing take-offs, landings, movement of aircraft, and the operations of all elevators. The air boss would direct air operations from here, having access to the navigation bridge for ship's course and speed information and tension controls of the arrestor wires (these had to be changed for different types of aircraft). He could also control the catapults, landing lights and wave off flights from here.

Finally there was a status board that showed the current weight and status of all aircraft. This board is filled in as if there is really a squadron in the sky/preparing to take off. There are a lot of photographs of flight operations that are in this room that the docents will show to guests when they tour this area. Of particular interest are the photos if landings using the safety net system. The view from this level of the ship and entire San Francisco Bay area is incredible.

02 Level

Back down to the 03 level, the ladder there can be taken to descend one more level to 02 and this is where the combat information center (CIC), air intercept center (AIC), carrier air traffic control center (CATCC), air operations (AOC), carrier-controlled approach (CCA), anti-submarine warfare command and control center (ASWCCC), and the radio central are located. These areas are not strictly in or under the island completely and actually spread out behind the Island under the flight deck. Currently only the CIC, AIC, ASWCCC, and radio central are accessible to the public and only by special tour.

CIC, AIC, ASWCCC

The CIC is where all the data about friendly and enemy craft is collected and then disseminated to where it is needed. The room has eight radar repeaters, potting boards, a dead reckoning trace, and a surface target plotter. The information here is constantly being updated with real-time information and the CIC is manned at all times. Friendly and enemy aircraft, surface craft and submarines are all tracked here on the various plotting boards. Each board has been filled with simulated or historical data as it would have really looked.

Connected to the CIC are the AIC and ASWCCC rooms with more radar repeaters, and displays showing various anti-submarine equipment used through the years (including a Russian sonobouy). The CIC is one of the most interesting parts of the ship and definitely worth the time to tour. The docents can turns the equipment on and the lights off so guests can have the true feeling of being in an operational CIC.

Left: Primary flight control at 08 level of island.

Below: Combat information center plotting boards on 02 level.

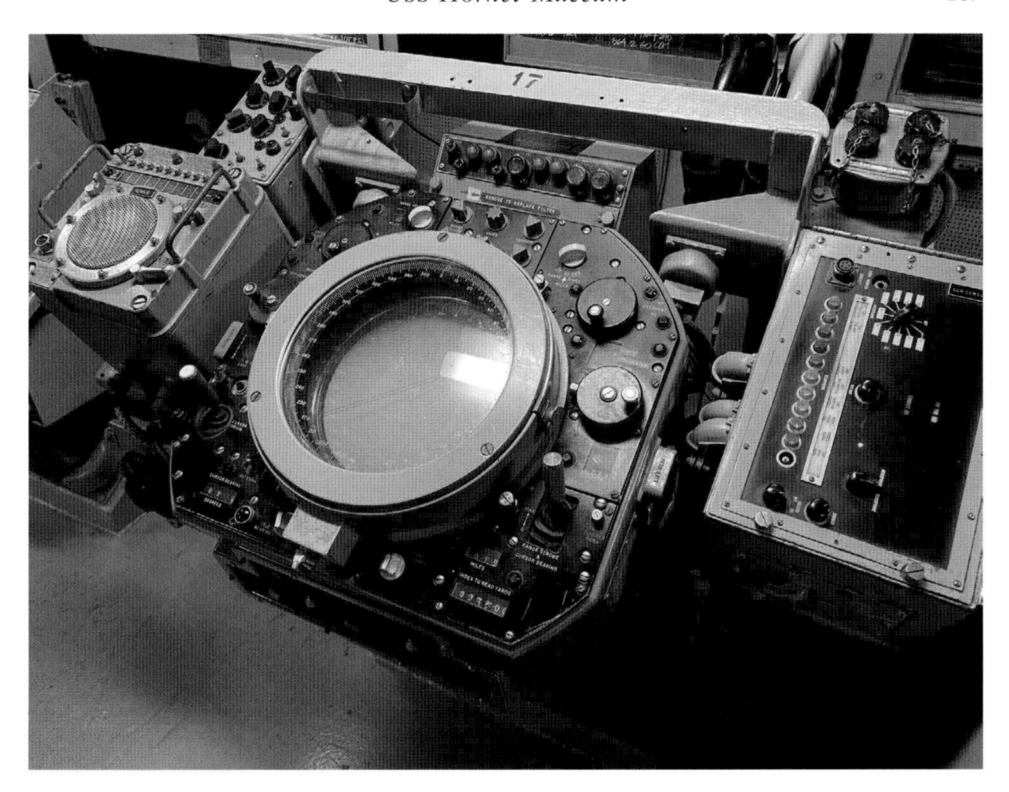

Combat information center radar repeater on 02 level.

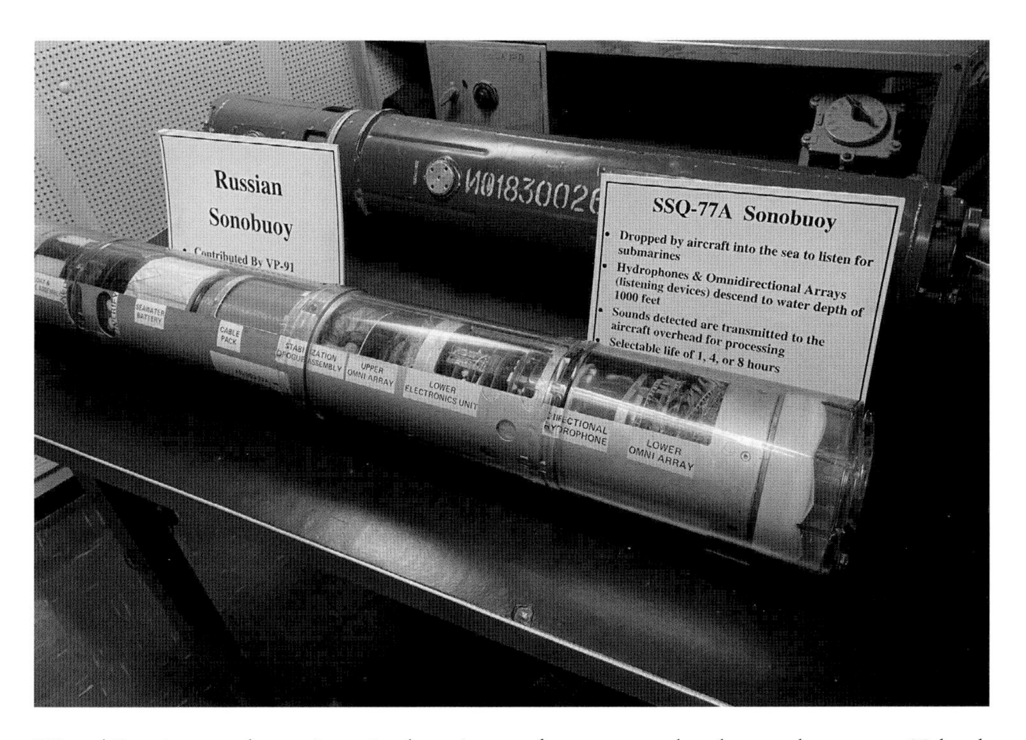

US and Russian sonobuoys in anti-submarine warfare command and control center on 02 level.

Radio Central

The USS *Hornet* amateur radio club operates out of several radio rooms. The first room has working HAM radios that are actively used by the club. They make over 3,000 contacts a year with other radio enthusiasts. They use donated US Air Force Harris RF 350 radios since the original radios were all removed or destroyed during the mothball era for the *Hornet*. These radios utilize the original radio antennas visible on the sides of the flight deck. Those interested can join the amateur radio club. Nearby are several rooms that make up the message processing center with the original radios, teletype machines and more. One of these rooms is the secret room where top secret messages would be decoded.

Second Deck (Starboard Side)

From any of the three hangar bays, there are many ladders down to the second deck. For ease of understanding, this description will start from the ladder on the starboard side in hangar bay one. There are numerous exhibits on the second deck. There are restrooms located right at the bottom of the first ladder.

Radio central on 02 level.

Officers' Store

A small officers' store holds vintage items that would have been available back when the ship was active. This includes Shinola shoe polish, Colgate tooth powder, Bayer aspirin, Lucky Strike cigarettes, "Oh Henry!" candy bars, and much more. On the walls here are two sets of controls here (with multiple levers and gauges) that allow the magazines located below to be flooded (first set) or drain the bilge tanks (second set). As guests go down the passageways on this level, they will see many controls for various units in other parts of the ship such as the water curtains in the hangar bay. It is easy to overlook these are there are so many but guests are encouraged to read the signs at each to try to get an idea of the workings of the ship.

Forward Auxiliary Emergency Generator

Proceeding forward in this passageway leads to the forward auxiliary emergency generator. This massive Fairbanks Morse diesel engine could generate 1000 kilowatts of power. It is so large that it can be seen from two different levels: the second deck and the hangar bay level (in the Fo'c'sle). It was originally designed as a train engine and looking at the size of it this it is not a surprise. There is a second generator in the back of the ship.

Fog/Foam Station

The next compartment holds a fog/foam station. This station was one of many found across the ship. The protein-based foam would be sprayed over fires, smothering them. The foam was more effective than water because fires could be fuel or oil based. There are 5-gallon cans of the foam concentrate in racks. These racks are all over the ship. Back then the foam was made using animal byproducts such as cow's blood.

Officers' Wardroom

Further down this passageway is the officers' wardroom/mess room and ready room two. The officers ate here and were served from the small adjacent pantry, although the food was not cooked there but brought up from below. The officers' pantry was open twenty-four hours a day. There are multiple historic photos of the *Hornet* here, as well as formal china dinnerware. Ready room two is connected to the wardroom (ready rooms are where pilots were briefed before missions) and has been converted into a classroom by the current *Hornet* educational staff.

Forward auxiliary power on second deck.

Fog/foam station on second deck.

Officers' wardroom on second deck.

Naval Aviation Exhibit

The room in between the wardroom and the galley has been turned into a special exhibit depicting the first aircraft landing on a ship. This room was originally the officers' lounge. There is a large model of the USS *Pennsylvania* showing the wooden deck built for the attempt. There is also a looping video that quickly explains the history of naval aviation. There is a large panel display of the history of aircraft carriers here, starting with the USS *Langley* (CV-1) and going all the way up to modern nuclear carriers. Eugene B. Ely landed his Curtis biplane on the deck of the *Pennsylvania* and his aircraft was stopped by lines attached to sandbags stretched across the deck. The lines caught on his improvised arrestor hook, which was (most likely), a meat-hook. This occurred on January 18, 1911 very close to where the *Hornet* is now in San Francisco harbor.

Officers' Staterooms

Next are two officers' staterooms that have been restored to their original state. These rooms show how tight the quarters were even for officers. Each room has two bunks, a sink and mirror, and a desk with a safe and a storage area. They also show how every nook and cranny was used for something in the ship. In most rooms the partitions between the rooms did not go all the way to the ceilings so there was not even full privacy for officers.

Air Group Eleven Exhibits

The next two rooms are also officers' quarters, but they have been converted to be combination exhibits and staterooms. They are dedicated to Air Group Eleven (from World War II). These two officer's staterooms contain large numbers of artifacts from members of Air Group Eleven donated to the museum. They also tell some of the stories Air Group Eleven airmen and one of the rooms features highlights from the movie *Eleven*. This movie is an interview of eleven members of Air Group Eleven by George Retelas whose grandfather served with them.

Executive Officer's Quarters

Next up is the restored executive officer's (XO) quarters. A small conference room holds historical ship and plane models around the conference table. Behind this is the XO's stateroom with a bed, desk, and its own bathroom (head). There is a very small desk outside where a Marine guard was always stationed. He would keep a log and report any incidents that needed to be recorded.

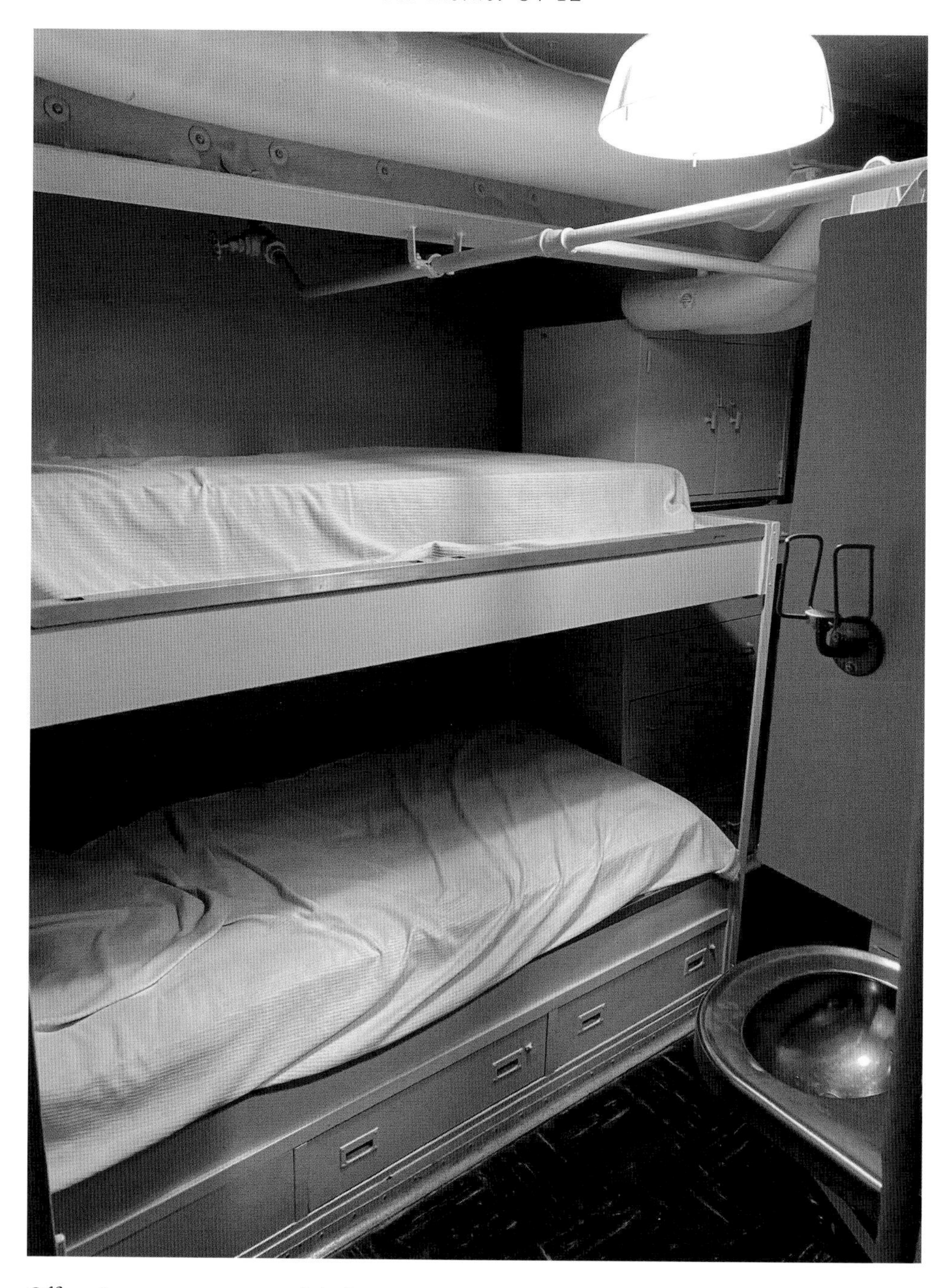

Officers' stateroom on second deck.

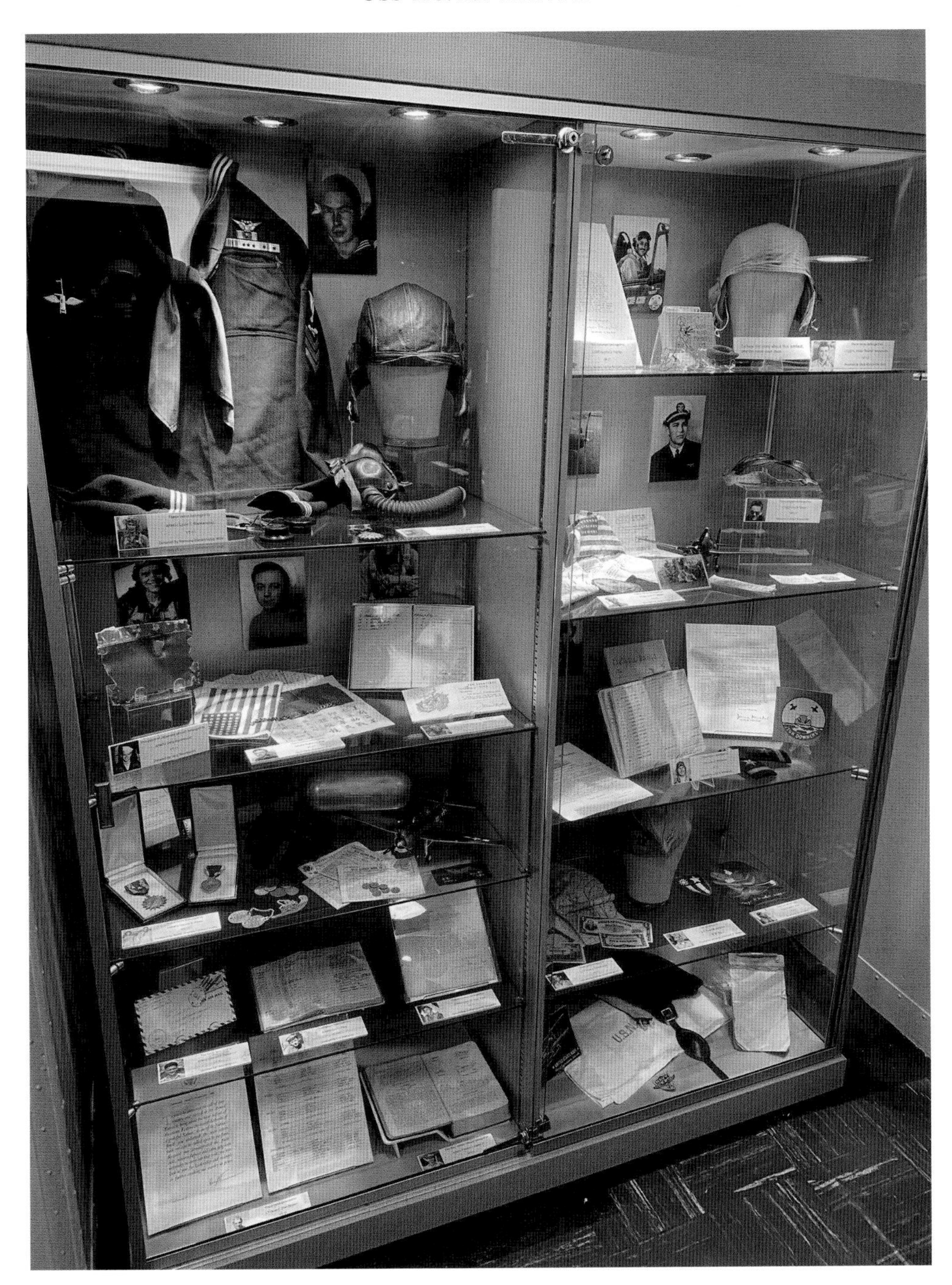

Air Group 11 exhibit on second deck.

Rotating Exhibit Area

The next room aft used to be ready room three. This room has been converted into a rotating exhibit area. There is information about how to maintain exhibits and artifacts and the rest of the room is used to show rotating exhibits. Each exhibit will typically last six months to a year and will feature items from the museum's artifact collection. A few of the former chairs from the ready room are around the sides. Historical pictures adorn the walls around the room.

Disbursing Office

The next area towards the stern is the administration area used by the staff and crew of the *Hornet*. This area holds the disbursing office where payroll was held and distributed, the ship's TV studio and more. This area is closed off to the public and contains offices, storage and meeting rooms for the employees and volunteers of the ship.

From Mothballs to Museum Exhibit

After this is the "From Mothballs to Museum" exhibit, that is housed in the Ray Vyeda Memorial Room. This exhibit tells the story of the ACHF outlined previously in this book. It has newspaper clippings of all the significant events that happened during the 1990s. Raymond Vyeda was one of the leaders of the efforts to save the ship and there is an article from the *San Francisco Chronicle* in 1995 that features him and the efforts to save the ship. Vyeda served on the *Hornet* in the 1950s. There is also a looping video, showing some of the significant moments. A quote from the mayor of Alameda is painted on the wall from 1995. In summary, it says that it would be impossible to save the *Hornet* and maintain it.

Anti-Submarine Warfare Exhibit

The next exhibit is the "Anti-Submarine Warfare" exhibit. This exhibit outlines the bow mounted sonar, the Anti-Submarine Classification and Analysis Center (ASCAC), and discusses the methods for searching for and attacking enemy submarines. There is also a nice display of sonobouys.

Ready Room Four

The next open room is the fully restored ready room four. This room has the original seats, maps, charts, some flight gear, and landing records on Air Group

Raymond Vyeda Memorial Room/From Mothball to Museum exhibit on second deck.

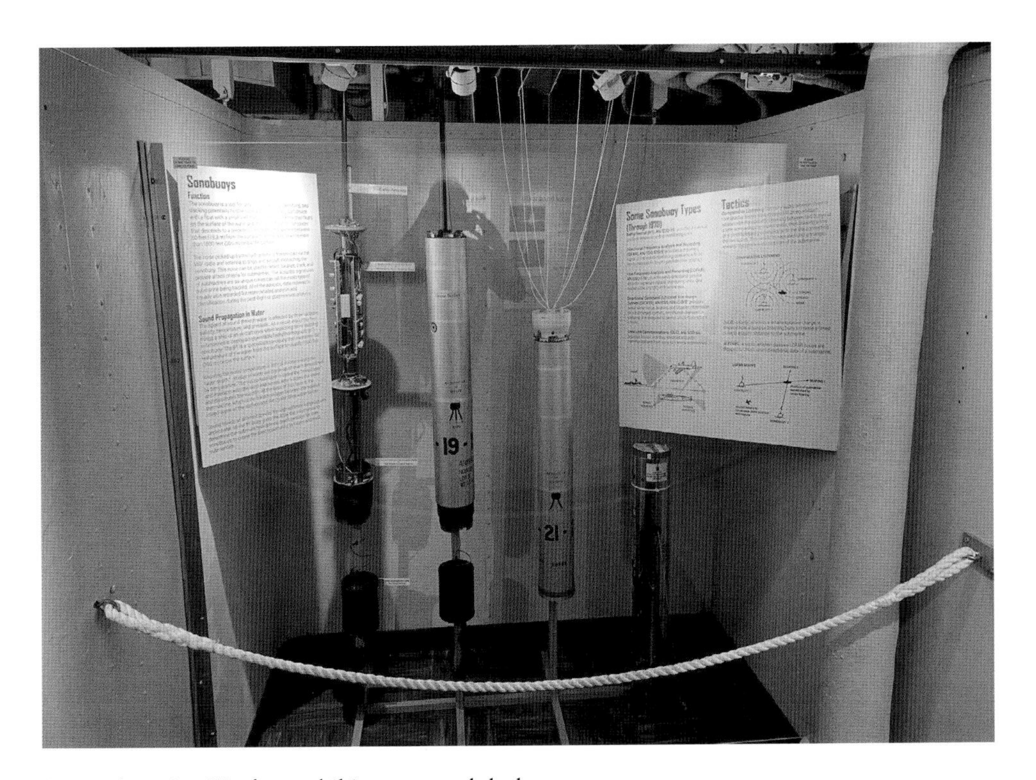

Anti-Submarine Warfare exhibit on second deck.

Ready room four on second deck.

VS-35, the Blue Wolves. This room was used to brief pilots before meetings, to have air-group meetings, and as a place for the pilots to rest and relax. The seats were the most comfortable on the ship and each could recline as well as having a small, foldable desk so the pilots could take notes. There is a binder near the entrance that contains stories told by actual pilots that served on the *Hornet*. If guests stand at the podium at the front they can almost feel like they are talking to a group if aircrew before a mission. Note that before the refits in the 1950s, the ready rooms were on the 02 level, under the flight deck but above the hangar bay. After some incidents in World War II where a single bomb that penetrated the flight deck killed an entire squadron of pilots in their ready room, they were moved down below the more heavily armored hangar bay. To compensate for the extra time it would take to get to the flight deck, an escalator was installed (right outside the door of ready room four) to bring the pilots to their aircraft quickly. This room is recognizable from the 2006 movie *Rescue Dawn* as the ready room that actor Christian Bale is in before his flight where he is shot down.

Marine Detachment Area

Further aft is the Marine Detachment (MARDET) area. This area contains the stores and pressing room, detachment office (used by the Marine Detachment commander), berthing, head, and their lounge. There are two berthing areas—one for the non-commissioned Marine officers and the other for the enlisted Marines. A large rack of restored M-1 rifles is in the passageway. In the lounge there is a large mural painted in the 1960s depicting an assault landing, vintage uniforms, and magazines. The Marine Detachment consisted of fifty-five enlisted men and non-commissioned officers, and two officers. It was their job to run the brig, guard the special weapons area, and to serve as the ship's police force. They helped on the anti-aircraft guns as well. They were typically the only crewmembers armed.

Post Office

Next is the ship's post office. This area is down a small hallway and in addition to the informational poster outside, you can partially enter the post office. Inside it is fully restored with mail slots (completer with letters), a desk, and countertop as well as a fully uniformed mannequin.

Garbage Disposal Room

Next is the garbage disposal room. This particular area was for food waste only. Two large receptacles are inside and food waste was brought here to be ground up before being flushed to the ocean through pipes. This was all according to navy regulations at the time.

Auxiliary Laundry

The auxiliary laundry room is next down the passageway. This particular laundry room was primarily used by the personnel from the sick bay and the Marine Detachment. It could also handle overflow laundry from the main laundry room. The room has a large washer, two dryers, as well as a heavy duty pressing (ironing) machine. The room was typically used twice a week and each department had its own day to use the room. After the laundry was collected and washed, it would be sorted (all clothes had the sailor's name stenciled on it) and then dropped off on their bunk. Each crewmember was responsible for putting their clothes away.

Marine Detachment Area with M-1 rifles on second deck.

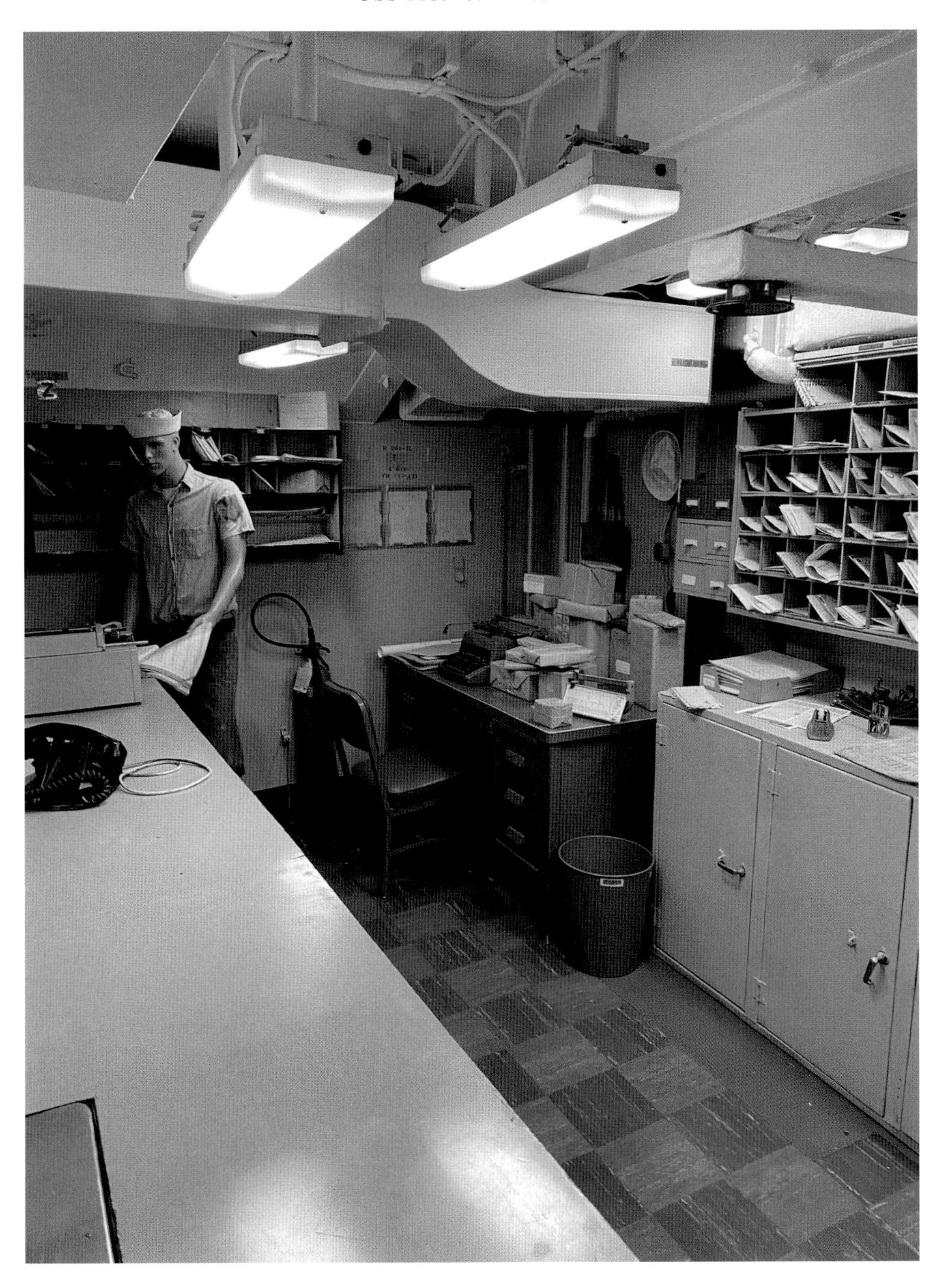

Ship's post office on second deck.

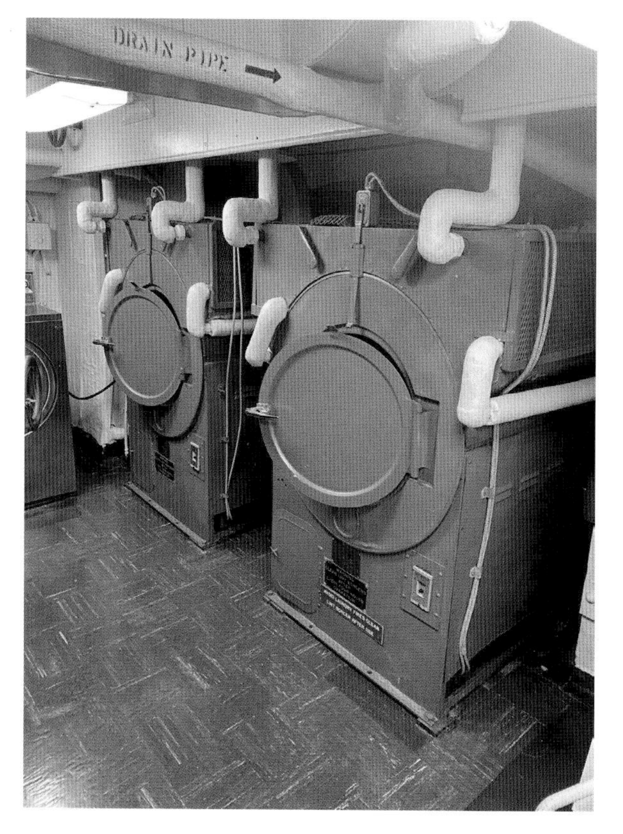

Above: Waste disposal room on second deck.

Left: Auxiliary laundry room on second deck.

Bluejacket Activity and *Hornet* Legacy Room

The next open area is actually two different areas connected to this passageway. The first is the *Hornet* Bluejacket activity room with tables, chairs, coloring books, and knot displays to keep the kids entertained. There is also the *Bluejackets' Manual Jr. Training Guide* on the wall that contains training information taken from an authentic 1946 *Bluejackets' Manual*. Across from that is a separate area showing vintage cartoons.

Beyond this is the *Hornet* legacy room where you will find the history of all seven of the previous ships named *Hornet* in the US Navy. This included a large Doolittle Raid exhibit. The first *Hornet* served during the Revolutionary War from 1775 until captured by the British in 1777. The second *Hornet* served from 1805–1806 in the First Barbary War where she helped the Marines capture Djerna and Tripoli. The third *Hornet*, which served from 1806–1829, was the first to be purpose built as a warship. This *Hornet* served with distinction in the War of 1812 and after, until lost in a storm. The fourth *Hornet* actually served at the same time as the third, from 1813–1820. The fifth *Hornet* was a steam powered side-wheeler that was captured from the Confederate Navy and served for the Union Navy in 1964. The sixth *Hornet* fought in the Spanish-American War in 1898. The seventh *Hornet* was the famous USS *Hornet* (CV-8) that participated in the Doolittle Raid, the Battle of Midway, and, finally, in the Battle of the Santa Cruz Islands where she was sunk. There are a lot of artifacts stored in this room, most donated by former crewmembers related to their time on the *Hornet*.

The *Hornet*'s legacy room on second deck.

Sick Bay

Further aft is the hospital/medical facility (sick bay) as the next opening off this passage. The whole sickbay is quite large with multiple operating rooms, an examination room, pharmacy, X-Ray lab, bacteriological room, sound room, head, offices, eye examination room, and a ward with comfortable bunks. In the ward, there is a section for seriously injured patients recovering from surgery, a quiet/isolation area, and the main area. A team of four MDs along with about fifteen medical corpsmen (who served as nurses) worked under a medical officer. There were a few pharmacist mates who ran the pharmacy. This team could handle minor doctor's visits all the way up to major surgeries. Oftentimes the *Hornet*'s sick bay would take care of sailors on the escort ships that had more limited facilities. Out to sea the ship would need to be able to treat any type of injury or malady so it had to be fully self-sufficient.

Sick bay ward on second deck.

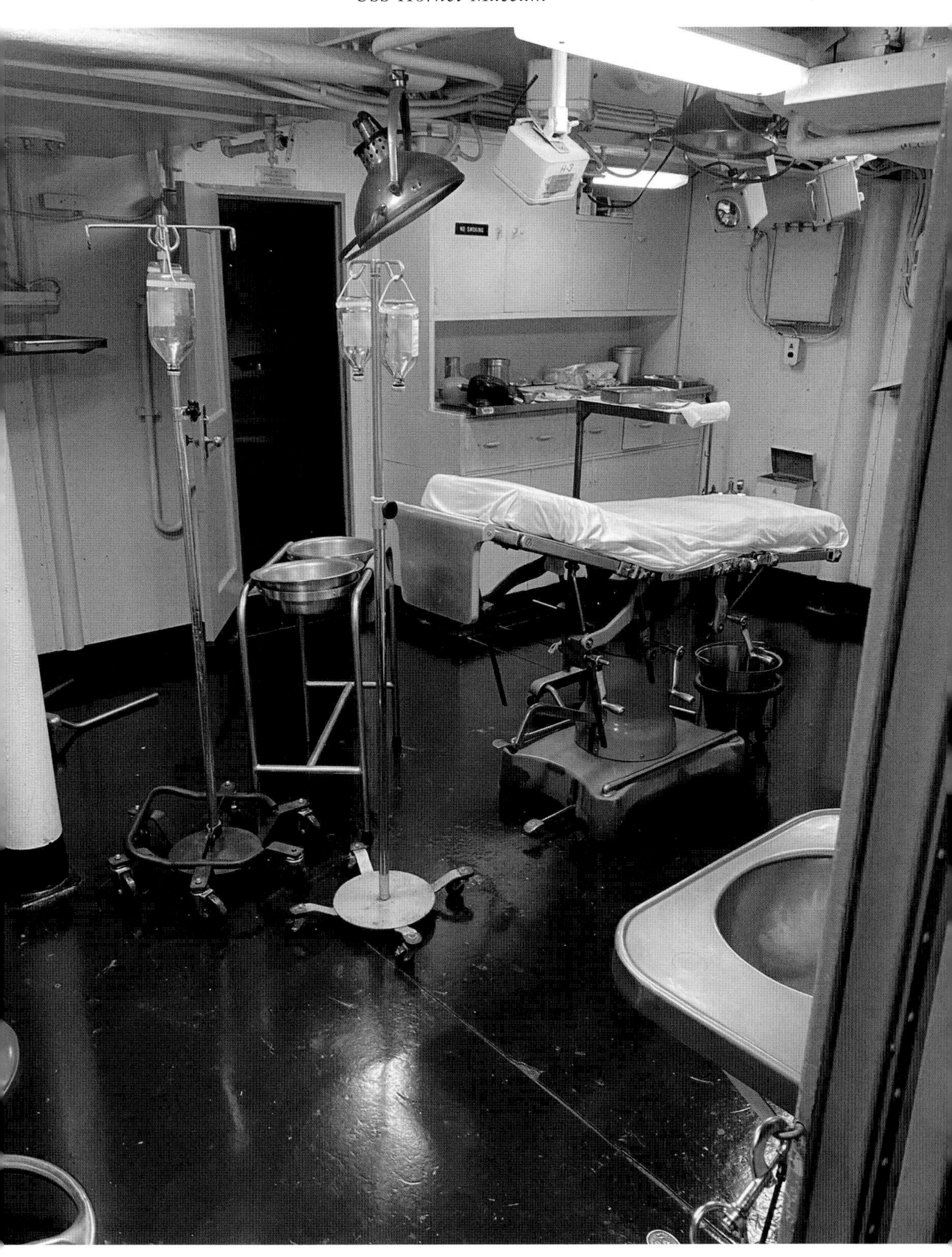

Sick bay operating room on second deck.

E Division

A door leads to E Division just outside of the sick bay. This is not an open area for guests to visit, but the placard describes the function of E Division which was the care and maintenance of the ship's electrical systems. They were also responsible for the ship's phones. They had shops throughout the ship to maintain all the equipment they were in charge of. Behind this door is their berthing area. Beyond this are other areas not open to the public, including the air conditioning office and the elevator machinery.

Women in the Military Exhibit

The next opening on the starboard side has two exhibits. The first one features the history of women in the military from the early 1900s until today. It has a timeline on the wall of significant events, a video with interviews of women in the military and mannequins wearing examples of uniforms through the years. Finally there is a small display of Rosie the Riveter-themed items.

African-Americans in the Military Exhibit

The second exhibit features the history of African-Americans in the military. It features a video timeline, biographies of notable African-Americans in US military history, a large map of segregated bases in the San Francisco Bay Area from World War II, and much more. This exhibit is maintained by the Walking Ghosts of Black History, a non-profit organization that focuses on bringing Black military history to life. Similar to the Friends and Family of Nisei Veterans, this exhibit is a partnership between the *Hornet* Museum and the Walking Ghosts of Black History.

Torpedo Workshop

Just beyond this is the torpedo workshop. To get there guests have to go through a small berthing area. The workshop has deep torpedo storage bins, examples of torpedoes from World War II up through the Vietnam War and the heavy worktables and overhead electric hoists (to move the torpedoes). Large railings on the ceilings here (and in other places on the ship where ammunition is moved around) facilitate the movement of heavy objects like torpedoes. An elevator in the rear of this shop would bring the torpedoes up to be placed on aircraft when ready. There is a lot of information on the walls about the different types of torpedoes used from the 1940s until 1970.

Right: Women in the Military exhibit on second deck.

Below: Torpedo workshop on second deck.

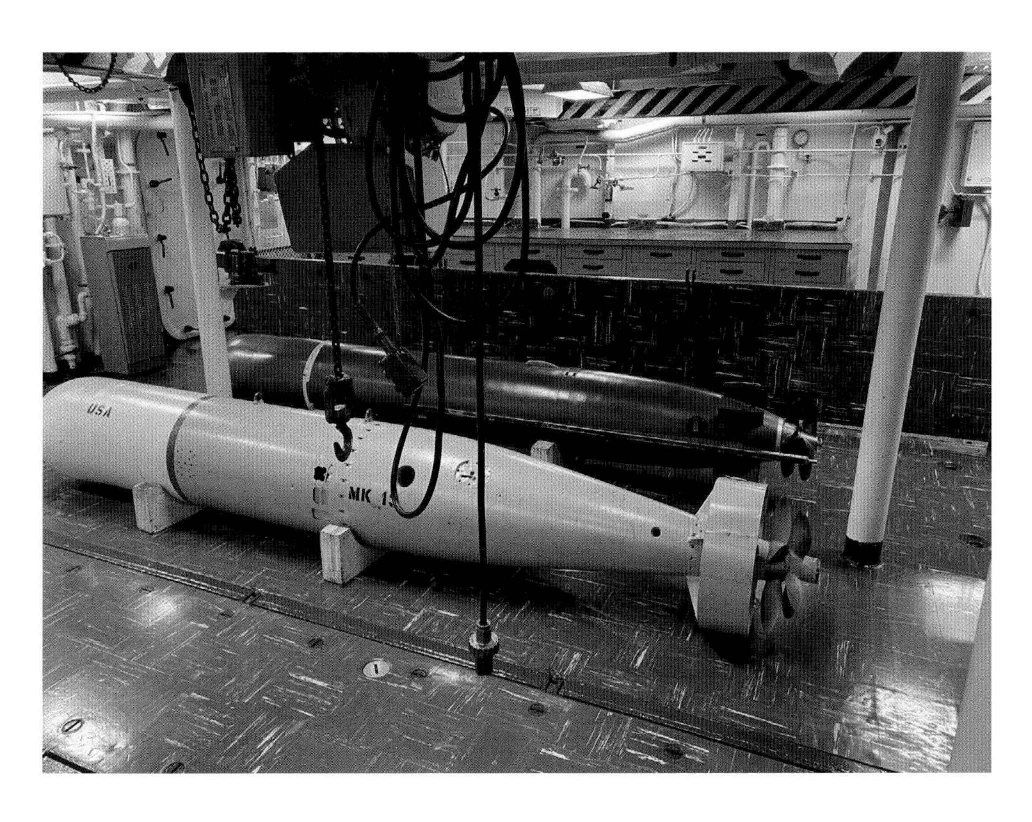

Ship's Chapel

Across from the torpedo workshop is the ship's chapel. The chapel has a large worship area with pews and a podium, chaplain's office, and a large library of historical books. The chapel would host small services by faith while larger services would be held in the hangar bay or on the flight deck. There was also a guest bathroom here for use when the ship was in port. For example if a woman needed to use the restroom back when the entire crew was male. The library was for the sailors' use. They could come here to read or to check out books. The collection in the library right now is a very respectable military collection. The museum crew has hung a large plaque in the chapel that lists all of the crewmen who lost their lives during their service time aboard the *Hornet.*

CPO Pantry, Mess, and Lounge

The last stop on the starboard side is the chief petty officers' (CPOs) pantry and mess. There is a small pantry here and tables and chairs for the CPOs to eat. Next door is the CPO lounge. The rudder centering device is located here in case the rudder is damaged it can be straightened here. This allows the ship to be steered by propellers alone. The ladder from here leads up to the aircraft maintenance

Ship's chapel on second deck.

area in hangar bay three. There is a last passageway leading aft through CPO berthing, the repair locker, and a machinery room. After this guests are at the rear of the ship and a ladder leads up to the fantail from here.

Second Deck (Port Side)

On the port side of the second deck, starting about halfway in hangar bay one and continuing all the way aft to the fantail, is another set of exhibits. These exhibits occupy a lot of spaces that were previously berthing areas for the crew.

Treasures from the Collection

The first exhibit occupies multiple compartments and is an art gallery of *Hornet* and navy paintings from the museum's collections called "Treasures from the Collection." This is an evolving exhibit and will continue to change over time.

Fire Safety Exhibit

The next open area has an exhibit about fire safety aboard the *Hornet.* Fire safety is taken very seriously by the navy and all personnel are trained in firefighting. There is a display here of some of the firefighting equipment that would be used to fight fires.

Task Force Exhibit

This is quickly followed by a still developing task force exhibit. The task force exhibit has diagrams depicting the standard formations of ships in the World War II fast carrier forces. There are also models representing the various ships types (aircraft carriers, light aircraft carriers, battleships, cruisers, and destroyers).

Artificial Reef Exhibit

The next area contains an exhibit describing how ships, including *Essex*-class carriers, whether sunk by accident, enemy action or deliberately sunk in specific locations can create artificial reefs. Information panels describe what artificial reefs are, effects on the environment and has examples of several that are near to the San Francisco Bay area. The USS *Macon*, an airship, crashed into the ocean in 1935 and was discovered in 1991 in 1,500 feet of water in the Monterey Bay National Marine Sanctuary. The fleet tug USS *Conestoga* was lost in 1921 and was recently discovered in 2014 near the Farallones Islands. The USS *Independence* (CVL-22), after surviving a nuclear bomb test at Bikini Atoll, was purposefully sunk off of the Farallones Islands and was also recently rediscovered in 2015. Although there was an attempt to decontaminate the ship, it has been found to still be radioactive.

Sister-Ship Row

Aft of that, filling multiple compartments is sister-ship row. This exhibit is extremely large and trying to describe it does not give a full picture. Every one of the other *Essex*-class carriers is featured with different aspects of their history highlighted. This area is filled with photographs, maps, artifacts and scale models. Although none of the *Essex*-class carriers were sunk in combat, they had many varying fates at the end of their service lives. Four are currently ship museums in various places around the US (in addition to the *Hornet* there is the *Yorktown*, *Intrepid*, and *Lexington*). Sister ship rooms 1 and 2 showcases the

Sister ship row deck crew vest display on second deck.

USS *Bennington* (CV-20), USS *Leyte* (CV-32), USS *Kearsarge* (CV-33), USS *Lake Champlain* (CV-39), USS *Tarawa* (CV-40), USS Princeton (CV-37), and the USS *Antiem* (CV-36). Sister ship room 3 showcases USS *Wasp* (CV-18), USS *Yorktown* (CV-10), USS *Bon Homme Richard* (CV-31) and USS *Valley Forge* (CV-45). Sister ship rooms 4–6 showcase USS *Franklin* (CV-13), USS *Philippine Sea* (CV-37), USS *Randolph* (CV-15), USS *Bunker Hill* (CV-17), USS *Essex* (CV-9), USS *Oriskany* (CV-34), USS *Intrepid* (CV-11), USS *Hancock* (CV-19), USS *Boxer* (CV-21), USS *Shangri-La* (CV-38), USS *Ticonderoga* (CV-14), and USS *Lexington* (CV-16).

Several small exhibits are scattered through sister-ship row, highlighting related aspects of the life of a sailor. There is also a repeat of the display explaining the vests worn by deck crew. A large size chart shows how the *Essex*-class compares to other aircraft carrier classes. A room in this area has been restored and filled with artifacts as if it was occupied by sailors in the 1960s. Another display displays the work done by the USO. Finally there is a small section that lets you test your communication skills with a few fun games.

Third Deck

Deeper in the ship there are many areas accessible only via docent led tours. This includes large portions of the third deck, parts of the fourth deck, the engine room, and the fire room. There are multiple ways to get to the third deck, but for ease of description, the third deck will be described back to front since that is more likely the way a guest's tour would go.

Enlisted Mess and Enlisted Galley

The rear-most compartment regularly included in tours is the enlisted mess and galley. The mess has dozens of metal tables and chairs in the large open room. There are also food stations around the edges-including a coffee machine. Right down the passageway from this is the galley. On your way to the galley is a large rack with metal trays for the food. Inside the galley are large cooking appliances of all types: ovens, kettles, grills, etc. Everything is bright stainless steel and gives you an idea of the large food operation that had to feed thousands of sailors four times a day. There were three normal serving times and a night-time option for late shifts, hence the four servings per day. At the front of the galley is the serving area. The sailors would come down the passageway with their trays and they could see the different types of food available behind the clear sneeze-guards. The walls of the passageway leading here are covered with various recipes that the galley workers would follow for making the large portions of food they prepared. There are also a few pictures of sailors working in the galley.

Enlisted mess on third deck.

Enlisted galley on third deck.

Machine Shop

The main machine shop of the ship is located next to the galley. This machine shop is still used by the restoration crew today and is equipped with a lot of original machine tools—although not from the *Hornet*. The tools here were recovered from other mothballed ships, but they are the same kinds of tools that would have been in the shop. Most major repair work was done in this shop, and if it could not be done here, then it most likely needed to be fixed in a shipyard. Right next to this room is the lid to one of the boiler (fire) room escape shuttles. Further up is the electrical shop and electrical division office, although you cannot see much inside of these rooms. They are still in use by the ship's restoration crew just like the machine shop.

This room has:

Five metal engine lathes.
One four-speed drill press.
One 36-inch radial drill press.
One horizontal milling machine.
One shaper machine.
One reciprocating saw.
One pedestal bench grinder.
One miller arc welder.
One 3-foot by 5-foot precision table.
One 50-ton shop press.

Bomb Elevators

Many of the bomb elevators open to this deck and three can be seen during this tour. The first is right outside of the enlisted mess. This elevator still operates and is often used to bring heavy equipment either up or down from both the hangar bay and the flight deck. This is the same one that guests would have seen outside the Apollo exhibit on the hangar bay and/or aft of the island on the flight deck. The other two are further forward and includes the largest one that was made to be large enough to bring up nuclear weapons if needed and is located next to the special weapons area.

Ge-Dunk

The Ge-Dunk (listed as "Crew's Messing" on the wall sign) is a place where the enlisted sailors could go in between normal meals to buy snacks, hot dogs, ice cream, coffee, and sodas and more. It is essentially a ship's cafe. There is a small galley here with coffee machines, soda fountain, ice cream machine, etc. Round tables are in the seating area and they have small signs advertising such things as hot dogs for 15 cents, root beer floats for 20 cents, etc.

Special Weapons Area

The special weapons area has a Marine guard post. This is where nuclear weapons were stored and they were secured and guarded by the Marines. Not everyone on board even knew that the ship carried nuclear weapons. This area is under restoration and will eventually be available for tours.

Bomb Assembly Area

Several rooms have been restored and are open for viewing if on a tour in this area. They include the bomb assembly area, the transfer passage, the bomb room, and the sidewinder magazine. This is just beyond the special weapons area and is one of the spots where the bomb elevators open to. The first area seen is the bomb assembly area and this has racks holding a 2,000-pound mine, a 100-pound practice bomb (filled with sand), and a hedgehog anti-submarine weapon. There are also pictures on the wall describing these weapons in greater detail. Next is the transfer passage, and in this room are dozens of bombs on pallets, some missile tubes, and some older-style practice bombs. The sidewinder magazine is not available to see at this time. The last room is the bomb room, and this holds one 500-pound bomb with its fins in the open position, one World War II-era 500-pound bomb, another 100-pound sand-filled practice bomb, and a 20-mm anti-aircraft gun and its storage box.

Catapult Machinery Room

The last area to be toured on the third deck is the catapult machinery room. This is the portside catapult and the room is full of some large machinery. This machinery includes the large cables, pulleys, hydraulic equipment, control panels, and more needed to catapult an aircraft on the flight deck into the air. There is over a mile of cable between the room and the catapult on the flight deck. The room has four high-pressure air tanks with clever names such as "Vodka" and "Rum" as well as one hydraulic fluid tank called "Beer." The crews obviously had a sense of humor.

Marine special weapons guard station on third deck.

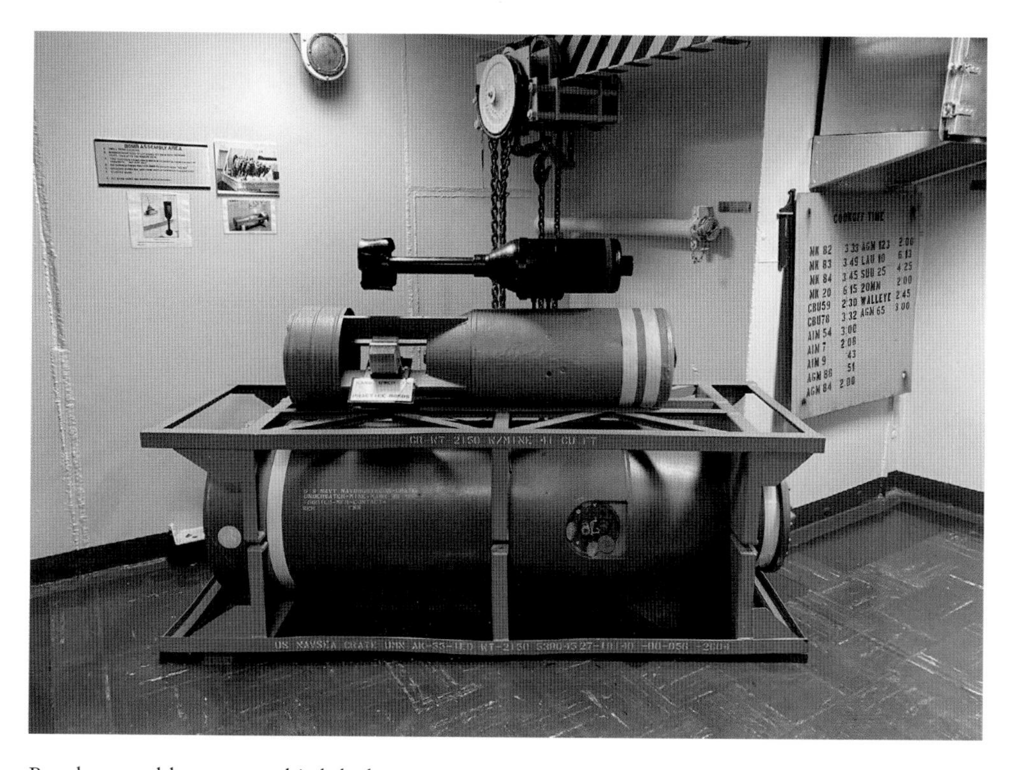

Bomb assembly area on third deck.

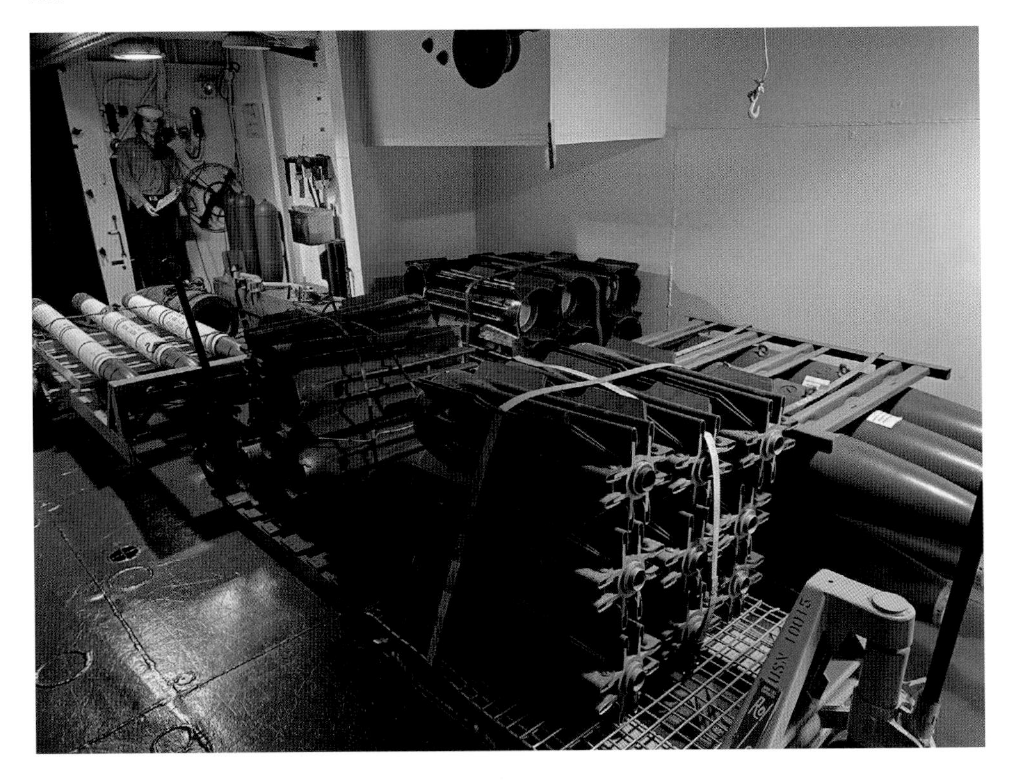

Above: Bombs and missile tubes stacked on pallets on third deck.

Left: Catapult cables in catapult machinery room on third deck.

Catapult hydraulic tanks in catapult machinery room on third deck.

Fourth Deck and Below

There are various ladders down from the third deck to the fourth deck and even further down to the engine rooms and the fire rooms. These areas are only accessible by tours with docents, and because of the nature of the ship's layout, the entrances to all these areas are fairly close to each other on the third deck. A single ladder leads to the tailor shop, cobbler shop, and the athletic gear locker.

Tailor (Fourth Deck)

The first door that is accessible opens to the tailor shop. This area has a large collection of authentic, donated uniforms hanging on a rack, two large presses, a sewing machine, a desk, and shelving.

Cobbler (Fourth Deck)

The second door opens to the cobbler shop. This area has racks of authentic, donated shoes, a shoe polishing and buffing machine, a desk, and some heavy duty sewing equipment. Behind the machines are some boxes of authentic Camel brand wire nails and Klean-Kutt nails and tacks.

Athletic Gear Locker (Fourth Deck)

The final door in this passageway opens to a fully outfitted athletic gear locker. When the crew had free time, whether on shore leave or even on board, they could come and check out sports equipment. The racks here hold bats, gloves, tennis rackets, basketballs, soccer balls, fencing equipment, golf clubs, boxing gloves, etc. There are vintage uniforms, trophies and other memorabilia here also. All of the artifacts here have been donated to the ship. There is a small office here as well.

Brig (Fourth Deck)

A stairway down from the special weapons area leads to the brig. There are several cells down here and they can be lit up from the inside to show what they were like. They were very small rooms and the doors have crisscrossing bars as you would expect. The most common use of these cells was for short-term disciplinary action against sailors.

Tailor shop on fourth deck.

Cobbler shop on fourth deck.

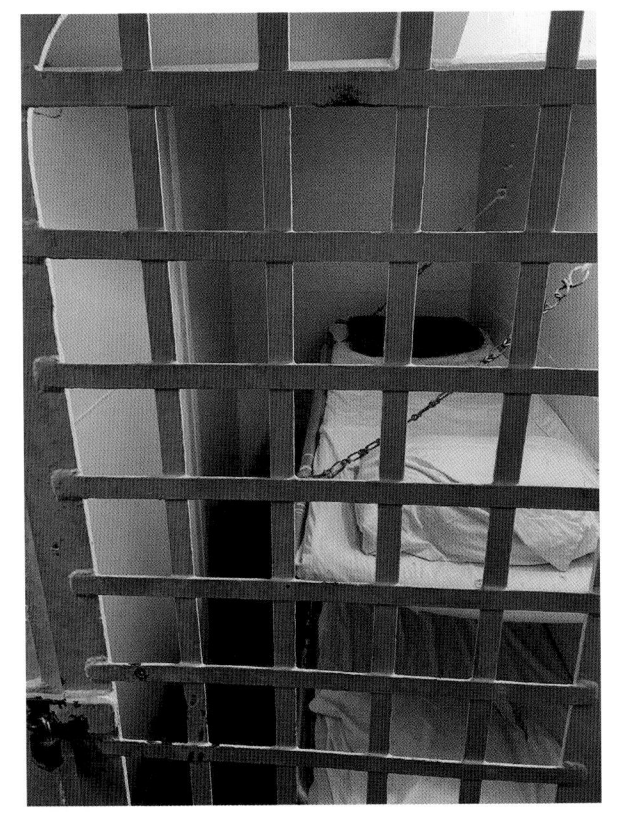

Above: Athletic gear locker on fourth deck.

Left: Ship's brig on fourth deck.

Engine Room (Fifth Deck)

After climbing down several ladders guests will reach engine room #1. This room houses the massive machinery needed to move a ship of this size. This includes the desalination tank, main condenser, de-aerating feed tank, booster pumps, low-pressure turbines, high-pressure turbines, reduction gears, 1250 KW turbo-generator, and more. This room operates the two outside propeller shafts. There are walls of dials and indicators here. This is truly one of the most interesting spots to see on the ship.

Fire Room (Fifth Deck)

Right next door to engine room #1 is fire room #1 although it is accessed by a separate ladder. In this room you will find the main feed pump, boilers, air vents, fuel oil heaters, and more. Like the engine room, this is truly one of the spots on the ship that guests should not miss. The entrance to the escape shuttle that opens near the mess area on the third deck above is here. There is a shaft with a long ladder leading up. In the event of an emergency where the regular ladders might not be accessible, crewmembers in the fire room could climb to safety from here.

ENGINE REVOLUTIONS
INDICATOR
SHAFT REVOLUTION INDICATOR
R.P.M.
KEEP OUT
AUTHORIZED
PERSONNEL ONLY
Engine Order Telegraph

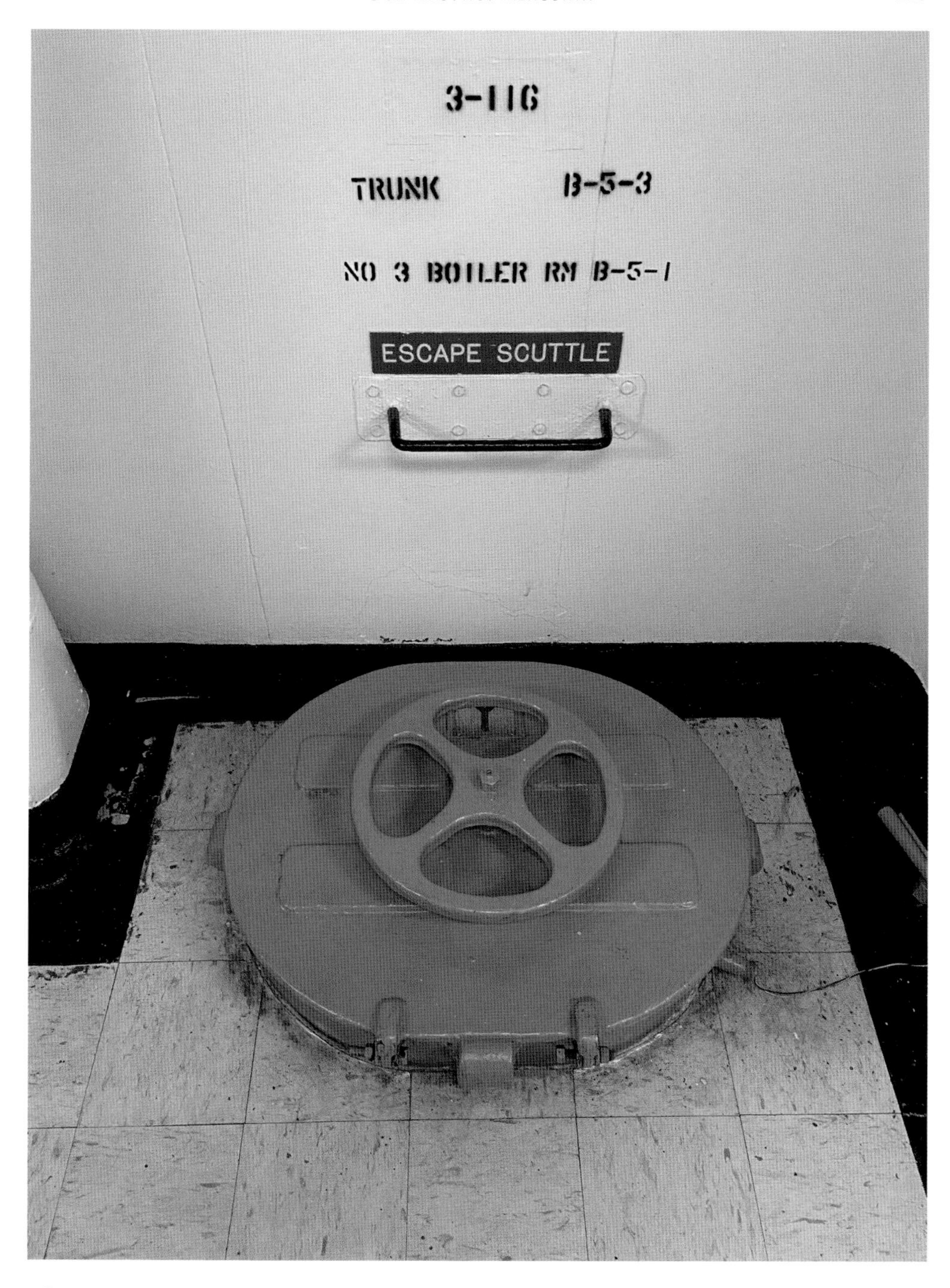

Above: Fire room escape scuttle hatch (on third deck).

Opposite above: Engine room 1 accessible from ladder on third deck.

Opposite below: Fire room 1 accessible from ladder on third deck.

8

Aircraft and Spacecraft on Display

The *Hornet* Museum has a large collection of aircraft and space program artifacts from all the eras she served in and beyond. World War II, Korea, Cold War, Vietnam, Gemini, Apollo, and modern aircraft are all represented. The range goes from piston engines to jet engines to helicopters to space capsules. Each exhibited aircraft (and large space program artifacts) are detailed.

FM-2 Wildcat

The Wildcat served from the Pearl Harbor attack until VJ Day in World War II and on all fronts. Pilots flying Wildcats shot down German, Italian, Vichy French, and Japanese aircraft during the war. The rugged fighter was often less maneuverable than its opponents but made up for this with toughness, dependability, and good armament. Even as new models such as the Hellcat and Corsair came along the Wildcat still served on escort carriers, training carriers, and on many naval air stations. Many US allies used them as well.

This particular FM-2 Wildcat (BuNo. 55052) was a training aircraft for aircraft carrier pilots in World War II. This training was carried out on Lake Michigan on two training carriers converted from civilian vessels. On June 14, 1945, Ensign E. J. Robinson was flying this aircraft off of USS *Wolverine* (IX-64) when he had engine failure and had to ditch in Lake Michigan. He escaped but the aircraft sank to the bottom of the lake. The Wildcat was recovered fifty years later in 1995 and turned over to the *Hornet* museum by the Naval Aviation Museum in Pensacola Florida in 2006 for restoration. The aircraft is on loan to the *Hornet* museum.

FM-2 Wildcat.

Specifications (FM-2 Wildcat)

Dimensions:	Length: 28 feet, 11 inches. Height: 9 feet, 11 inches. Wingspan: 38 feet, folded 14 feet, 4 inches. Wing Area: 260 square feet.
Weights:	Empty: 5,448 pounds. Loaded: 7,487 pounds. Maximum Take Off: 8,271 pounds.
Fuel:	Internal Load: 117 US gallons Maximum External Load: two 50 US gallons external tanks
Power plant:	One Wright R-1820-56/-56A or -56W/-58WA (with water injection) Cyclone 9 radial piston engine rated at 1,350 horsepower
Performance:	Maximum Speed at 28,000 feet: 332 miles per hour. Cruising Speed: 164 miles per hour. Climb Rate: 3,650 feet per minute. Service Ceiling: 34,700 feet.
Range:	Normal: 900 miles. Maximum: 1,310 miles with external fuel.
Armament:	Four wing-mounted 0.5-inch Browning M2 machine guns each with 430 rounds of ammunition, plus six under wing rocket projectiles or up to 500 pounds of bombs.
Crew:	One pilot.

TBM-3E Avenger

The Avenger (named because it was supposed to be unveiled on December 7, 1941 and this became a symbol of revenge against the Japanese sneak attack) torpedo bomber was the successor for the obsolete Devastator torpedo bombers the US Navy started World War II with. The plane had a crew of three and could carry a torpedo, 2,000 pounds of bombs, or depth charges. They could also carry rockets, but after some combat experience, the Avenger was found not to be a good rocket platform. It was faster, tougher, and better armed than its predecessor, and the Avenger stayed in service throughout the war. The combat debut of the Avenger was at the Battle of Midway where a squadron of six Avengers (detached from the *Hornet*'s ill-fated Torpedo Squadron 8) unsuccessfully attacked the Japanese fleet from Midway and lost five aircraft. However, the Avenger went on to be a very successful aircraft in the torpedo bomber, level bomber, and anti-submarine roles. Later versions included radar for night operations and many other variants. Many US allies used Avengers as well.

This particular aircraft was purchased by the museum in 1998 and restored with squadron markings from VT-17 that served on the *USS Hornet* in World War II. VT-17 helped to sink the super-battleship *Yamato* in 1945.

TBM-3E Avenger.

Specifications (TBF-1C)

Dimensions: Length: 40 feet.
Height: 16 feet, 5 inches.
Wingspan: 54 feet 2 inches, folded 18 feet.
Wing Area: 490 square feet.
Weights: Empty: 10,555 pounds.
Loaded: 16,412 pounds.
Maximum Take Off: 17,364 pounds.
Power plant: One 1,700-hp Wright R-2600-B Cyclone fourteen-cylinder two-row radial piston engine.
Performance: Maximum Speed at 12,000 feet: 257 miles per hour.
Cruising Speed: 153 miles per hour.
Climb Rate: 768 feet per minute.
Service Ceiling: 21,400 feet.
Range: With Torpedo: 1,105 miles.
As Scout: 2,335 miles.
Armament: Two 0.5-inch machine guns firing ahead (earlier models had .03-inch machine guns), one 0.5-inch machine gun in dorsal turret, and one 0.3-inch machine gun in lower rear position, one torpedo or up to 2,000 pounds of bombs, mines, or depth charges.
Crew: Three (pilot, radio operator, and bombardier).

T-28B Trojan

The T-28 was a training aircraft for the US Air Force, Navy, and Marine Corps. In the navy, the T-28 was in service from the 1950s until the 1980s. The aircraft also saw some combat in the Vietnam War and served in other countries before and since.

This particular aircraft has been painted in the standard yellow overall as a training aircraft. This model is an air force model painted in navy markings. The most obvious difference is the lack of an arresting hook (which would be the T-28C).

Specifications (T-28C)

Dimensions: Length: 33 feet.
Height: 12 feet, 8 inches.
Wingspan: 40 feet, 1 inch.
Wing Area: 268 square feet.
Weights: Empty: 6,424 pounds
Maximum Take Off: 8,500 pounds
Power plant: One Wright R-1820-86 Cyclone nine-cylinder air-cooled radial engine, 1,425 horsepower.

T28B Trojan.

Performance: Maximum Speed at 10,000 feet: 343 mph
Climb Rate: 3,540 feet per minute.
Service Ceiling: 35,500 feet.
Range: Normal: 1,060 miles.
Armament: Six hard points with a capacity of 1,200 pounds total
Crew: Two.

FJ-2 Fury

The FJ-2 Fury is the naval version of the F-86 Sabre and resulted from the need to quickly bring aircraft capable of fighting the MiG-15 onto the US carriers. Like the Sabre, the Fury was still a gun fighter, carrying four 20-mm cannons in the nose. The design was rushed and in the end all of its problems were never resolved. The biggest problem was its lack of low speed capabilities (needed to land on aircraft carriers). Despite modifying the Fury with folding wings, stronger landing gear, and other changes, the Fury never really transitioned into a carrier aircraft. They were instead given to the Marines who used them mostly from land bases. However, a few were received by the Marines before the end of the Korean War and they used them on carriers despite their issues.

This particular aircraft has been painted with the markings of VMF-312 "Checkerboards," a Marine squadron. They flew the Fury from 1954 until 1956. It is on loan from the National Museum of Naval Aviation in Pensacola, Florida.

FJ-2 Fury.

Specifications (FJ-2 Fury)

Dimensions:	Length: 37 feet, 7 inches.
	Height: 13 feet, 7 inches.
	Wingspan: 37 feet, 1.5 inches.
	Wing Area: 288 square feet.
Weights:	Empty: 11,802 pounds.
	Maximum Take Off: 18,790 pounds.
Power plant:	One General Electric J47-GE-2 Turbojet with 6,000 pounds standard thrust.
Performance:	Maximum Speed: 675 miles per hour.
	Climb Rate: 7,230 feet per minute.
	Service Ceiling: 46,800 feet.
Range:	Normal: 860 miles.
Armament:	Four 20-mm Colt Mk. 12 cannons.
Crew:	One pilot.

F-11 Tiger Cockpit

The F-11 was a navy fighter from 1956 until 1961. The Tiger was the successor to the Cougar by Grumman. After 1961, it was mostly used as a trainer as it was quickly superseded in performance by the F-8U Crusader. The F-11 was the

F11 Tiger Cockpit.

second operational navy aircraft to be supersonic. The F-11 was used by the Blue Angels from 1957 until 1968 when it was replaced by the Phantom. A funny story about the F-11 is that it is considered the first jet aircraft to shoot itself down. Pilot Tom Attridge was flying an F-11 in 1956 and test fired the 20-mm cannons while diving. Somehow his trajectory and those of the shells crossed paths and his aircraft was damaged enough to force him to make a crash landing. He survived, but his aircraft did not.

This particular aircraft cockpit (BuNo. 141821) entered service in 1958 and served in multiple squadrons until retired in 1964. It was obtained by the museum from the Pima Air and Space Museum in 2003.

Specifications (F-11 Tiger)

Dimensions: Length: 46 feet, 11 inches.
Height: 13 feet, 3 inches.
Wingspan: 31 feet, 7.5 inches.
Wing Area: 250 square feet.

Weights: Empty: 13,810 pounds.
Loaded: 21,035 pounds
Maximum Take Off: 23,459 pounds

Power plant: One Wright J65-W-18 Turbojet, 7,400 pounds standard thrust (10,500 with afterburner).

Performance: Maximum Speed: Mach 1.1

Cruising Speed: 577 miles per hour.
Climb Rate: 16,300 feet per minute.
Service Ceiling: 49,000 feet.

Range: Normal: 1,275 miles.

Armament: Four 20-mm Colt Mk. 12 cannons, four hard points for AIM-9 Sidewinders, bombs, or fuel tanks

Crew: One pilot

TA-4J Skyhawk

The TA-4J was the training version of the A-4 Skyhawk. It had two seats and was used to train carrier pilots. The TA-49 first entered service in 1969 and stayed as the primary navy jet trainer until the 1990s. It has served in multiple roles since then, most notably as an "adversary" squadron aircraft where its high maneuverability makes it a tough opponent. This was depicted in the movie *Top Gun* when the trainers were in Skyhawks.

This particular aircraft was in service until 2003 (as part of VC-8) when it was flown to Oakland and then towed to the USS *Hornet* museum. It is on loan from the National Museum of Naval Aviation in Pensacola, Florida.

TA-4J Skyhawk.

Specifications (TA-4J Skyhawk)

Dimensions:	Length: 43 feet, 7 inches. Height: 15 feet. Wingspan: 27 feet, 6 inches. Wing Area: 260 square feet.
Weights:	Empty: 12,273 pounds. Maximum Take Off: 24,500 pounds.
Power plant:	Pratt & Whitney J52-P6A with 8,500 pounds standard thrust.
Performance:	Maximum Speed: 660 miles per hour. Cruising Speed: 587 miles per hour. Climb Rate: 8,440 feet per minute. Service Ceiling: 42, 250 feet.
Range:	1,956 miles.
Armament:	Two Colt, 20-mm, Mk. 12 cannons and up to 7,900 lb. of ordnance carried externally on three hard points.
Crew:	Two pilots

F-8U-1 Crusader (plus Cockpit)

The F-8 Crusader has often been called the last of the gunfighters because it still retained its cannon armament after most other new fighter aircraft were only being equipped with air-to-air missiles for fighting other aircraft. The F-8 was the first supersonic navy fighter and was primarily used in the Vietnam era. The F-8 was the most successful fighter aircraft the US used in Vietnam in terms of air-to-air kill ratios. This could be attributed to the aircraft and the superbly trained pilots.

This particular aircraft is on loan from the National Museum of Naval Aviation in Pensacola Florida. Originally built in 1957, the aircraft was restored by *Hornet* volunteers and carries the squadron markings of the first F-8 unit operational in the Seventh (Pacific) Fleet: VF-154 serving on the *USS Hancock* (CVA-19) starting in 1958. The *Hornet* also has an F-8 cockpit that can be opened to let people sit inside. This cockpit is mobile and can towed to locations off ship for community events.

Specifications (F-8E Crusader)

Dimensions:	Length: 44 feet, 6 inches. Height: 15 feet, 9 inches. Wingspan: 35 feet, 2 inches. Wing Area: 350 square feet.
Weights:	Empty: 17,541 pounds. Combat: 25,098 pounds. Maximum Take Off: 34,000 pounds.

F-8U 1 Crusader.

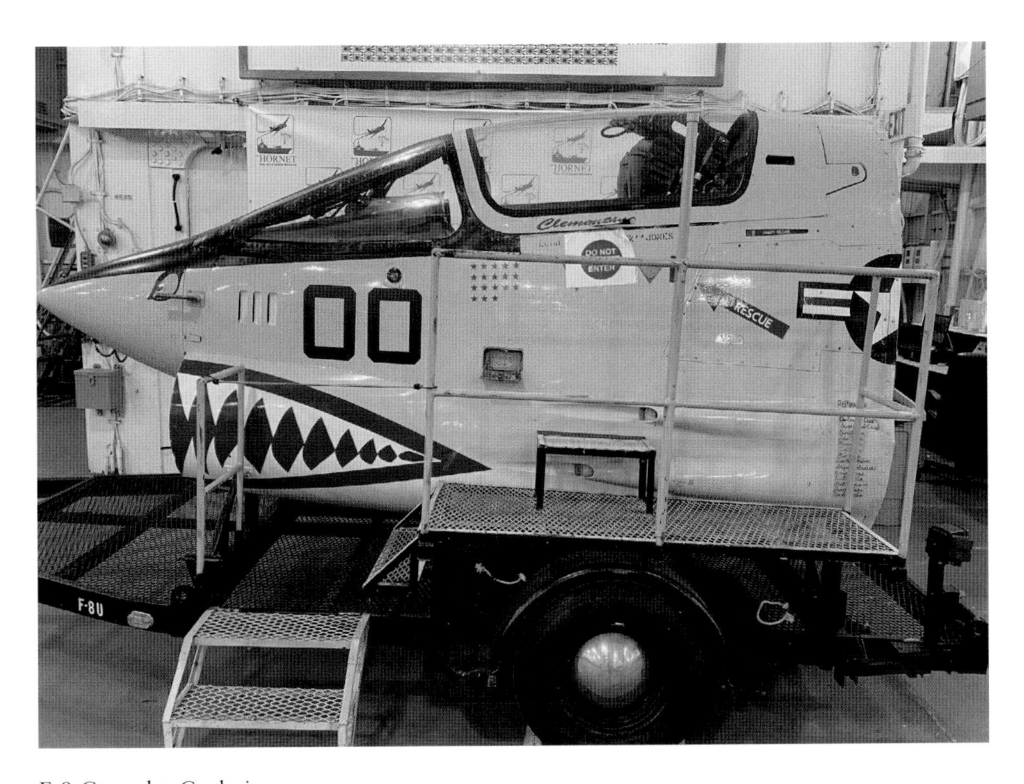

F-8 Crusader Cockpit.

Power plant: One Pratt & Whitney J57-P-20A turbojet rated at 10,700 pounds static thrust or 18,000 pounds standard thrust with afterburner

Performance: Maximum Speed at 40,000 feet: 1,120 miles per hour.
Cruising Speed: 570 miles per hour.
Climb Rate: 31,950 feet per minute.
Service Ceiling: 58,000 feet.

Range: Normal: 453 miles.
Maximum: 1,737 miles.
Armament: Four Colt-Browning (20-mm) Mk. 12 cannon with 144 rounds per gun, plus up to four AIM-9 Sidewinder AAMs, or twelve 250-lb bombs or eight 500-lb bombs or eight Zuni rockets or two AGM-12A or AGM 12B Bullpup A AGMs.

Crew: One Pilot

US-2B Tracker

The S-2 Tracker was the first US anti-submarine aircraft designed to both detect and attack enemy submarines. They first entered service in 1952 and were active until 1970. Many navies and civilian groups (such as firefighters) still use them today. For detection, the aircraft could deploy sonobouys and use on board systems like the AN/APS-38 radar and the magnetic anomaly detector (MAD) AN/ASQ-8 locate and identify enemy submarines. It also had a powerful searchlight (70 million

US-2B Tracker.

candlepower). Once locating an enemy submarine, the Tacker could attack with nuclear or conventional depth charges, torpedoes, or multiple types of rockets.

This particular aircraft (BuNo. 136691) was donated to the *Hornet* in June 2000. It has been recently repainted as an aircraft from VS-35 to honor the loss of an S-2 Tracker and their crew during the Vietnam War in 1966 (as detailed in this book's Vietnam War section).

Specifications (S-2E)

Dimensions:	Length: 43 feet, 6 inches.
	Height: 16 feet, 7 inches.
	Wingspan: 72 feet, 7 inches. Folded 27 feet, 4 inches.
	Wing Area: 496 square feet.
Weights:	Empty: 18,750 pounds.
	Loaded: 24,413 pounds.
	Maximum Take Off: 29,150 pounds
Fuel and Load:	Internal Load: 4,368 pounds.
	Maximum ordnance: 4,810 pounds.
Power plant:	Two Wright R-1820-82 WA Cyclones each rated at 1,525 horsepower.
Performance:	Maximum Speed at sea level: 265 miles per hour.
	Cruising Speed: 207 miles per hour.
	Patrol Speed at 1,500 feet: 150 miles per hour.
	Climb Rate: 1,390 feet per minute.
	Service Ceiling: 21,000 feet.
Endurance:	8 hours.
Armament:	Early versions could carry one Mk. 34 or one Mk. 41 or two Mk. 43 torpedoes or one Mk. 24 mine in the bomb bay and either four Mk. 19 mines or four Mk. 43 torpedoes or four Mk. 54 depth charges or six HVAR rockets under the wings. Later versions had the bomb bay enlarged to carry nuclear weapons.
Crew:	Four (pilot, co-pilot/navigator, magnetic anomaly detector operator, other sensory equipment operator)

F-4J Phantom II

The F-4 Phantom is perhaps the best-known aircraft of the Vietnam War. The aircraft was used by the US Air Force, US Navy, and Marines. In the 1960s, the F-4 Phantom was a big, tough, fast aircraft and over 5,000 were built. Originally constructed as a missile only fighter, lessons learned in the Vietnam War resulted in the early Phantoms being modified with a gun pod (20mm cannon) and later fighters having them built in. Most fighter aircraft since have been designed with cannons built in. Phantoms served from 1960 until 1996.

F-4J Phantom II (photo in service, currently in restoration).

This particular aircraft (BuNo. 153879) was delivered as an F-4J and later converted to an F4-S. This aircraft was the last Phantom to launch from an aircraft carrier, the USS *Midway* (CV-41) in 1986. The aircraft has been restored to its unit markings from when it retired. This aircraft was left at NAS North Island until loaded on the USS *BonHomme Richard* (LHA-6) and transported to San Francisco for Fleet Week and then brought over to the *Hornet* on a barge. The aircraft is on long-term loan from the National Museum of Naval Aviation in Pensacola, Florida.

Specifications (F-4E Phantom II)

Dimensions:	Length: 63 feet.
	Height: 16 feet, 5.5 inches.
	Wingspan: 38 feet, 7.5 inches, folded 27 feet, 7 inches.
	Wing Area: 530 square feet.
Weights:	Empty: 30,328 pounds.
	Loaded: 41, 487 pounds.
	Maximum Take Off: 61,795 pounds.
Fuel:	Internal Load: 1,855 gallons
	Maximum External Load: One 600-gal. tank centerline, two wing-mounted 370-gallon tanks
Power plant:	Two General Electric J79-GE-17A turbojets, each rated at 17,900 pounds thrust with afterburning.

Performance:	Maximum Speed: Mach 2.2. Climb Rate: 41,000 feet per minute. Service Ceiling: 62,250 feet.
Range:	Normal: 786 miles. Maximum: 1,978 miles.
Armament:	Fixed internal M61A1 Vulcan 20-mm six-barreled cannon (earlier versions had a cannon pod or no cannon), four AIM-7 Sparrow missiles, and four AIM-9 Sidewinders. Hard points allow for wide variety of ordnance to be carried including bombs, gun pods, ECM pods, nuclear weapons, etc.
Crew:	Two (pilot, navigator/sensor operator).

S-3B Viking

The S-3 was designed to be the replacement for the S-2 and entered service in 1974. The S-3 was an improvement on the S-2 in every way and was very successful. During its career, it had multiple roles, from ASW to anti-ship to refueling aircraft, electronic warfare, and more. One big improvement was the centrally integrated data computer for all the different sensor systems. This allowed a crew of four on the S-3 to function similarly to the P-3 Orion with its crew of twelve. The S-3 served through the Gulf Wars and only was retired from service at the end of 2015.

This particular aircraft, model BuNo. 160599, was first assigned to VS-31 at NAS Jacksonville, Florida. The squadron was transferred to the USS *John*

S-3B Viking.

C Stennis (CVN-74). In August 2007, after participating in Operations Iraqi Freedom and Enduring Freedom, this aircraft was retired. Like the F-4 Phantom, the S-3 was left at NAS North Island and was loaded on the USS *BonHomme Richard* (LHA-6) and transported to San Francisco for Fleet Week and then brought over to the *Hornet* on a barge. The aircraft is on long term loan from the National Museum of Naval Aviation in Pensacola, Florida.

Specifications (S-3A Viking)

Dimensions:	Length: 53 feet, 4 inches. Height: 22 feet, 9 inches. Wingspan: 68 feet, 8 inches. Wing Area: 598 square feet.
Weights:	Empty: 26,650 pounds. Loaded: 42,500 pounds. Maximum Take Off: 52,540 pounds.
Fuel:	Internal Load: 12, 863 pounds Maximum External Load: Up to two 300-gallon tanks
Power plant:	Two General Electric TF34-GE-2 turbofans each rated at 9,275 pounds standard thrust.
Performance:	Maximum Speed Sea Level: 506 miles per hour. Cruising Speed: 403 miles per hour. Climb Rate: 4,200 feet per minute. Service Ceiling: 35,000 feet.
Range:	Normal: 1,088 mils. Endurance: 7 hours, 30 minutes.
Armament:	Four Mk. 46 or Mk. 50 torpedoes, four Mk. 36, 62, or 82 bomb/destructors, or two B57 nuclear depth charges in the two weapon bays. Two wing pylons allow for additional weapons including Mk. 52, 55, 56, or 60 mines, rocket pods, decoys, AGM-84 Harpoons, or AGM-84E SLAM missiles.
Crew:	Four (pilot, co-pilot/tactical coordinator, tactical coordinator, sensor operator)

F-14A Tomcat

The Tomcat was developed by Grumman with all the lessons from the Vietnam War. The first flight was in 1970 and it became the first variable-wing air-superiority fighter. It was built with the range, power, ceiling, speed, maneuverability, and weaponry to fight all airborne threats. The Tomcat became famous for appearing in the movie *Top Gun* with Tom Cruise and Val Kilmer at the controls. The Tomcat was too large to serve on older carriers like the *Hornet*.

F-14A Tomcat.

This particular aircraft (BuNo. 162689) was built in 1986, served with several squadrons and fought in Operation Desert Storm. When it was taken out of service in 2002, it was flown directly to San Francisco International Airport and then brought to the *Hornet*. Like many of the museum's aircraft, it is on loan to the museum from the National Museum of Naval Aviation in Pensacola, Florida.

Specifications (F-14A Tomcat)

Dimensions:	Length: 62 feet, 8 inches. Height: 16 feet. Wingspan: 64 feet, 1 inch. Wing Area: 565 square feet.
Weights:	Empty: 40,104 pounds. Loaded: 59,714 to 70,764 pounds. Maximum Take Off: 72,000 pounds.
Fuel:	Internal Load: 2,385 gallons. Maximum External Load: Two 267-gallon external tanks.
Power plant:	Two Pratt & Whitney TF30-P-412A/414A turbojets each rated at 20,900 pounds standard thrust
Performance:	Maximum Speed: 1,544 miles per hour. Cruising Speed: 633 miles per hour. Climb Rate: 30,000 feet per minute. Service Ceiling: 50,000 feet.
Range:	Radius: 197 miles. Maximum: 2,000 miles.
Armament:	One M61A1 Vulcan Cannon. Standard is four AIM-54s, two AIM-7s, two AIM 9s. Can carry a vast array of weapons including bombs, laser guided bombs, rocket pods, ECM pods, etc.
Crew:	Two (pilot, radar intercept officer)

HUP-1 Retriever

The HUP-1 Retriever was a utility helicopter that first became operational in 1949 and stayed in service until 1958 (1964 in limited use). They were eventually carried on most aircraft carriers of the time and performed search and rescue, transportation, and other general duties.

Designed by Piasecki Helicopter, the Retriever had dual, overlapping three-bladed rotors that could be folded for storage making it very easy to store on carriers. Of note is the Retriever was the first helicopter to perform a loop, although this was unintentional. They served in combat through the Korean War and overall performed very well.

This particular Retriever (BU124915) was built in 1948 and was ultimately donated to the museum by a former HUP pilot Steve Linsenmeyer and is painted with the markings of a HUP-1 that served on the *Hornet* from 1954–1958.

Specifications (HUP-1)

Dimensions:	Length: 32 feet.
	Rotor Diameter: 35 feet.
Weights:	Empty: 4,100 pounds.
	Useful Load: 1,650 pounds.
Power plant:	550-horsepower Continental R975-42 Radial Engine.
Performance:	Maximum Speed: 108 miles per hour.
	Cruising Speed: 84 miles per hour.
	Service Ceiling: 10,200 feet.
Range:	Normal: 360 miles.
Armament:	Typically none.
Crew:	Two crew and four to seven passengers.

UH-34D SeaHorse

The Sikorsky H-34 was a very successful helicopter design that was ultimately in service from 1954 until the 1970s. The H-34 was also known as the Choctaw, SH-34 (Seabat), and UH-34 (Seahorse). It was used in the transport, cargo carrier, search and rescue, anti-submarine, and even briefly in the gunship role. This helicopter saw extensive use in the Vietnam War and was the last piston-engined helicopter used by the Marine Corps.

The H-34 served in the US Army, US Navy, Marine Corps, Coast Guard, and in many allied nations. It was considered low-maintenance and was especially highly regarded by the Marines. The H-34 debuted in the Algerian War with the French who pioneered many of the uses of helicopter to insert and withdraw troops from combat. In Vietnam, it was heavily used by the Marines and well over 100

HUP-1 Retriever.

UH-34D Seahorse.

were lost in combat and accidents (this number illustrates their heavy use rather than any deficiencies). This model also participated in the Cuban Missile Crisis, withdrew troops from the Dominican Republic and recovered astronauts in the Mercury program. Most famously, a Marine UH034D recovered Alan Shepard, the first US astronaut in space.

This particular model (BuNo. 150553) served in Vietnam from land bases and on an aircraft carrier. It was damaged seven times by enemy fire. When retired from active service, it was in storage in Arizona starting in 1971 until it was sold in 1983 and then sold again in 1986 before ending up at the Pima Air Museum. Pima Air Museum then donated it to the USS *Hornet* Museum in 2003 where it was restored with the markings of its service in Vietnam as part of HMM-363.

Specifications (UH 34D)

Dimensions:	Length: 46 feet, 9 inches.
	Height: 15 feet, 11 inches.
	Width: 5 feet, 8 inches.
	Rotor Diameter: 56 feet.
Weights:	Empty: 7,900 pounds.
	Gross Weight: 13,000 pounds.
Power plant:	One Wright 1820-84, nine-cylinder, radial, air-cooled 1,525-horsepower engine.
Performance:	Maximum Speed: 123 miles per hour.
	Climb Rate: 1,100 feet per minute.
	Service Ceiling: 9,500 feet.
Range:	Normal: 225 miles.
Armament:	Typically none.
Crew:	pilot, co-pilot, crew chief plus twelve troops or eight stretchers.

SH-2F Seasprite

The SH-2 (later called the UH-2) had a long career with the US Navy. First introduced in 1962 as a fast, utility helicopter, the SH-2 served until 1993. During that time, its role included anti-submarine, search and rescue, medical evacuation, cargo, reconnaissance, and even gunship. Kaman Aircraft Corporation submitted their design in response to the navy's request for a new, all-weather, compact, multi-purpose helicopter. The navy selected the Kaman model and production started. Initial models were under-powered and later ones were built with two engines.

The SH-2 served during the Vietnam War primarily in search and rescue mode. As other helicopters began to enter service the SH-2 seemed destined for an early exit from their service life. However, their compact size meant some smaller ships (such as frigates) could only operate the SH-2 and thus, with modifications, their service life was extended. They even served in Operation Desert Storm in 1991.

This particular SH-2F Seasprite (BuNo. 149021) was accepted into navy service in 1961 as a HU2K-1. Starting in 1962, she served on aircraft carriers and was redesignated as a UH-2A. In 1971 she was converted to a UH-2D (search and rescue) helicopter and served in that capacity until 1974. At that point, she was converted to the SH-2F version for use as an anti-submarine helicopter. Finally retiring in 1991, she was given to the USS *Hornet* Museum on a long-term loan basis from the National Museum of Naval Aviation in Pensacola, Florida.

Specifications (SH-2F Seasprite)

Dimensions:	Length: 38 feet, 4 inches. Height: 13 feet, 7 inches. Rotor Diameter: 44 feet.
Weights:	Empty: 7,040 pounds. Maximum Take Off: 12,500 pounds.
Fuel:	Internal Load: 396 gallons. External Load: 120-gallon tanks.
Power plant:	Two General Electric T58-GE-8F turbo shaft engines, 1,350 standard horsepower each.
Performance:	Maximum Speed: 165 miles per hour. Cruising Speed: 150 miles per hour. Climb Rate: 2,440 feet per minute. Service Ceiling: 22,500 feet.
Range:	Maximum Fuel Load: 422 miles.
Armament:	2 side fuselage mounting stub/pylon stations carrying two Mk. 46 or 50 ASW torpedoes.
Crew:	Three (pilot, co-plot/tactical coordinator, sensor operator).

SH3H Sea King

The Sikorsky Sea King is one of the more successful US helicopters. Originally built as an anti-submarine helicopter, it was built like the S-2 Tracker to both find and attack submarines. It was the first all-weather helicopter in the US Navy. The hull is watertight and it can land on water with its landing gear retracted. At the time, it was the largest amphibious helicopter in service. The twin engines were seen as a major safety feature due to its ability to fly with one engine if needed. The Sea King first entered service in 1961.

During its long service life, Sea Kings have fulfilled almost every role possible for a helicopter: anti-submarine warfare, transportation, anti-shipping, medevac, airborne early warning, search and rescue, combat support, presidential transport,

SH3H Sea King.

and astronaut recovery. In Vietnam, they were used primarily to rescue downed pilots, either near their carriers or in enemy territory. The Sea King was the first helicopter to fly completely across the US, launching from the *Hornet* off the coast of San Diego, California, and flying all the way to Jacksonville, Florida, to land on the *USS Roosevelt (CVA-42).*

Some Sea Kings are still in service today. In the US, they are often used as firefighting aircraft and are called "Fire-Kings."

This particular Sea King (BuNo. 148999) started her service in the navy in 1961. In 1965, this helicopter recovered the crew of Gemini 4 from the USS *Wasp* (CVS-18). In 1975, she was modified into the SH-3H and served until 1994. Her next "mission" was to serve as the recovery helicopter in the movie *Apollo 13*. After this she was transferred to the "bone-yard" in Tucson, Arizona, until 2005 when she was given to the USS *Hornet* Museum as a permanent loan by the National Museum of Naval Aviation in Pensacola, Florida. She has retained the paintjob she had for the *Apollo 13* movie.

Specifications (SH3H)

Dimensions: Length: 54 feet, 9 inches.
Height: 16 feet, 10 inches.
Rotor Diameter: 62 feet.
Weights: Empty: 11, 856 pounds
Loaded: 18,626 pounds.
Maximum Take Off: 21,500 pounds.
Power plant: Two General Electric T58-GE-1o turbo shaft engines with 1,400 standard horsepower each.
Performance: Maximum Speed: 166 miles per hour.
Cruising Speed: 136 miles per hour.
Climb Rate: 2,200' feet per minute.
Service Ceiling: 14,700 feet.
Range: Normal: 625 miles.
Armament: Up to 840 lb. of weapons total. Typical packages included two Mk. 46/44 anti-submarine torpedoes, sonobuoys, or a nuclear depth charge.
Crew: Four (Pilot, co-pilot, two sensor operators).

Apollo Block 1 Command Module CM-011

CM-011 was developed and constructed as a test module to simulate a command module with astronauts that could be sent into space. These were designed and built by Rockwell of Downey, California. They were developed before the Moon program was fully articulated so they were designed without the ability to dock or to let crew in and out via a pressurized chamber.

CM-011 was deployed in the AS-202 Mission described elsewhere in this book. That mission tested all the basic features of the module: structural integrity, stage separation, control systems, heat shield, and more. It was mostly a complete module, except it was not built to actually hold astronauts and that extra space was used for more recording equipment.

After the *Hornet* successfully recovered CM-011 in 1966, Rockwell conducted a land impact test by dropping the module from a crane. The large crack in the exterior that resulted from that test can still be seen today. This crack explains why all subsequent missions continued to use water landings.

This module is on loan to the museum from the Smithsonian National Air and Space Museum. It was given to the museum to restore and display.

Dimensions: Length: 36 feet, 2 inches.
Diameter: 12 feet, 8 inches.
Mass: Dry: 26,300 pounds.
Launch: 32,390 pounds (Earth Orbit).

Apollo Block 1 command module CM-01.

Payload:	2,320 pounds.
Performance:	Powered by: One AJ10-137.
	Maximum Thrust: 91.19kN.
	Burn Time: 750 seconds.
Crew:	Three (no crew carried on AS-202).

Mobile Quarantine Facility (MQF)

The MQF is not technically a spacecraft, but it is a large artifact parked in the hangar bay and was an important part of the Apollo missions so it is included here. The one on display in the museum is from Apollo 14 and is serial number 004. Astronauts Alan Shepard, Edgar Mitchell, and Stuart Roosa from Apollo 14 all spent time in this MQF. The MQF was designed to allow returning astronauts and a small extra crew to stay in quarantine in relative comfort while on board their recovery ship.

The MQF is a converted Airstream trailer built to be completely self-contained with advanced air ventilation and filtration systems. The wheels were removed and instead the trailer was installed on a 35-foot-long base that strengthened the whole structure and allowed for easier transportation. It has six bunks, a

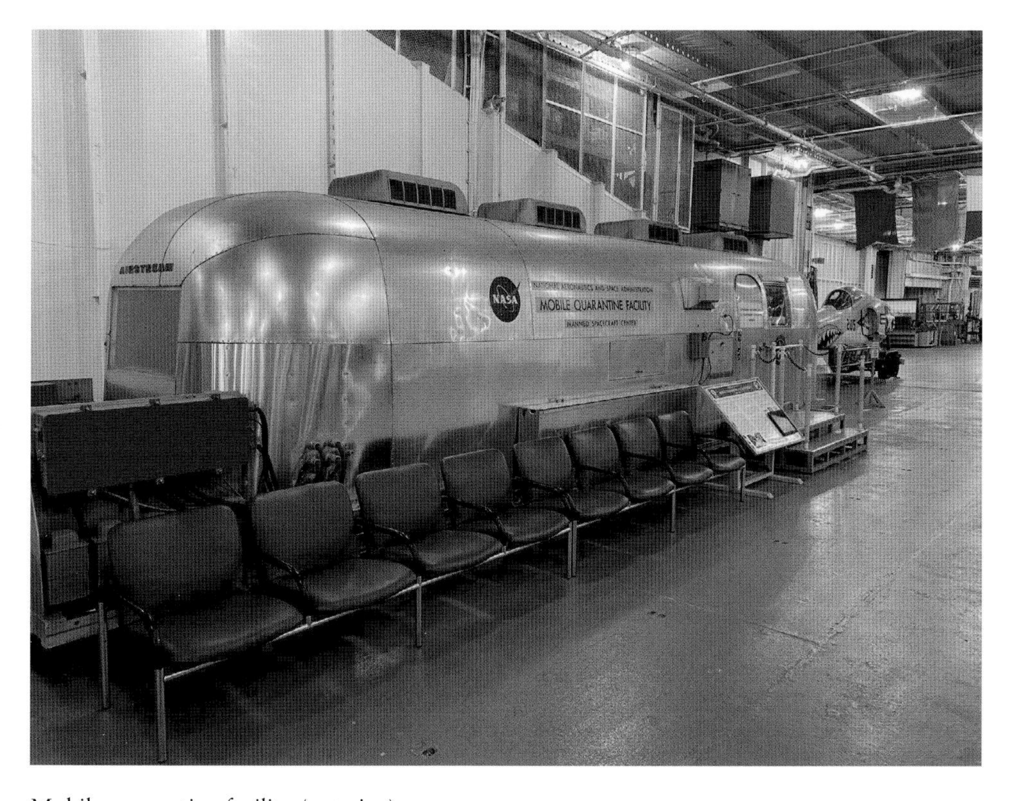

Mobile quarantine facility (exterior).

lavatory, a galley, a small lounge (with six seats and removable table), control panel for in-facility systems, and extensive communications equipment. There was a transfer lock that allowed food to be brought in and lunar samples to be passed out safely. It was also built to be easily transportable so it could be moved on and off of a ship and also airlifted in large cargo planes.

This MQF was donated to the museum by the Kansas Cosmosphere in Hutchinson Kansas in 2001.

Dimensions:	Length: 35 feet.
	Height: 8 feet, 7 inches.
	Width: 9 feet.
Weights:	Empty: 11,856 pounds.
	Loaded: 18,626 pounds.
Crew:	Typically three astronauts plus one doctor and one engineer.

Gemini Capsule Biolerplate (MSC-307)

The Gemini Capsule Biolerplate (MSC-307) is also not technically an actual spacecraft but it was built and designed to help in training for space missions. Biolerplates are used to test developing designs before they are finalized. This lets you test parameters like overall size and shape, aerodynamics, etc. They are also used in training for such things as maintenance, entry and exit, lifting and moving, connections to other systems, and more. It was likely used in training for the splashdown and recovery Gemini program.

MSC-307 is on loan from the W Foundation, a privately funded non-profit founded by Ken Winans and Debbie Wreyford-Winans in 2002. The W Foundation is "A 501c3 nonprofit organization dedicated to public education on the history of space exploration and financial literacy through public exhibits of their collection." MSC-307 was purchased from Colfax, California where it was residing as a park ornament for some time. In 2005, the W Foundation loaned it to the USS *Hornet* Museum where it was fully restored.

Dimensions:	Height (Length): 9 feet, 6 inches.
	Diameter: 7 feet, 6 inches.
Weight:	3,500 pounds.
Crew:	Two

Gemini boilerplate MSC-307.

Bibliography

Advisory Council on Historic Preservation/Protecting Historic Properties. achp.gov/protecting-historic-properties

Airstream/History Spotlight: Airstream's Place in Space, airstream.com/blog/history-spotlight-airstreams-place-in-space/

Alameda Sun/*Hornet*'s Last Commander Passes Capt. William M Pardee, alamedasun.com/obituaries/4880

Ballenger, W. and deFreitas, D., *USS Hornet Chronological Pictorial History: CV8 - CV-12 Volume I and Volume II.* (Independently Published, 2021)

Ballenger, W. and deFreitas, D., *USS Hornet Chronological Pictorial History: CVA-12 - CVS-12 Volume III and Volume IV.* (Independently Published, 2021)

Chinalakealumni/Capt.HollisterNewStationCommander, chinalakealumni.org/downloads/rocketeer/1957/rktr09.06.1957.pdf

Chinalakealumni/Capt. Hardy to be ComNOTS; Capt. Yount Future NAF CO, chinalakealumni.org/Downloads/Rocketeer/1964/Rktr03.27.1964.pdf

Colt, D., HistoryLink.org/Russell, Admiral James Sargent (1903-1996), historylink.org/File/10412

UCR Centers for Biographical Research/ RAdm Brandley in Council Race, cdnc.ucr.edu/?a=d&d=CJ19680222.2.18&e=-------en--20--1--txt-txIN--------1%25253E

Faltum, A., *The Essex Aircraft Carriers* (Charlston: The Nautical & Aviation Publishing Company of American. 1996)

Doud, B., Coronado Eagle & Journal/Coronado's "Avenue Of The Heroes" ... Captain Hoyt D., coronadonewsca.com/news/coronado_city_news/coronado-s-avenue-of-the-heroes-captain-hoyt-d-mann-usn/article_a08de18c-051f-11e8-9082-c30b04a60880.html

Eden, P. and Moeng, S., ed., Aircraft Anatomy A Technical Guide to Military *Aircraft from World War II to the Modern Day* (London: Amber Books. 2018)

Fish B., *Hornet Plus Three.* (Reno: Creative Minds Press. 2009)

Fold3 by Ancestry/Act Rep, Wadke-Sawar-Sarmi Area, New Guinea, 4/21 – 4/24/44, fold3.com/image/274205116
Fold3 by Ancestry/Rep Air Ops Against Luzon Is, Philippines on 11/5 & 6/44, fold3.com/image/292509272
Fold3 by Ancestry/Rep of air ops Against Formosa, Philippines, French Indo-China, So China & Ryuku Is, 1/3 – 22/45, fold3.com/image/295209547
Fold3 by Ancestry/Rep of Air Ops against Luzon Is, Philippines, 11/16-20/44, fold3.com/image/292508189
Fold3 by Ancestry/Rep of Air Ops Against Luzon Is, in Support of Landings on Mindoro Is, P 1 12/10 - 23/44, fold3.com/image/293473682
Fold3 by Ancestry/Rep of Air Ops Against the Ryuku Is, Formosa & the Philippines, Including Engagement with Jap Fleet, 10/2 – 27/44 & AA Acts on 10/13, 14 & 15/44, fold3.com/image/292246858
Fold3 by Ancestry/Rep of Air Ops in the Visayan Is, Area & Against Central Luzon Is, Philippines 11/7 – 15/44, fold3.com/image/292510534
Fold3 by Ancestry/Rep of Ops against the Palau & Philippine Is & in the Support of Landings on Morotai, Period 9/7 – 24/44, including AA action on 9/22/44, fold3.com/image/287146464
Fold3 by Ancestry/Rep of Ops Against Truk & Ponape Is, 4/29/44 – 5/1/44, fold3.com/image/274206613
Fold3 by Ancestry/Report of air operations against Japan, Ryukus & Jap Task Force, 3/18/45 – 4/27/45, fold3.com/image/295845758
Fold3 by Ancestry/Report of air operations against Kyushu and Shikoku Japan and the Ryuku Is, 5/12 – 27/45, fold3.com/image/296128314
Fold3 by Ancestry/Report of air operations against Kyushu, Japan and the Ryuku Islands, 5/28/45 – 6/10/45 with ACA Reports, fold3.com/image/296110356
Fold3 by Ancestry/Report of air ops against Japan, Bonins & Ryukus, 2/16/45 – 3/1/45, fold3.com/image/295352637
Fold3 by Ancestry/Report of Air strikes against the Marianas and Bonin Islands and action against the Jap Fleet during the period 11-24 June 1944, fold3.com/image/279824364
Fold3 by Ancestry/Report of Operations, Period 7/3/44 to 8/5/44 – Air Strikes on Bonin, Marianas and Caroline Islands, fold3.com/image/280043725
Fold3 by Ancestry/War Diary, 1/1 - 31/44, fold3.com/image/270951181
Fold3 by Ancestry/War Diary, 1/1 – 31/45, fold3.com/image/295252010
Fold3 by Ancestry/War Diary, 2/1 – 28/45, fold3.com/image/295365970
Fold3 by Ancestry/ War Diary, 2/1 – 29/44, fold3.com/image/274203577
Fold3 by Ancestry/ War Diary, 3/1 – 31/44, fold3.com/image/279779638
Fold3 by Ancestry/ War Diary, 3/1 – 31/45, fold3.com/image/295918451
Fold3 by Ancestry/ War Diary, 4/1 – 30/44, fold3.com/image/274533764
Fold3 by Ancestry/ War Diary, 4/1 – 30/45, fold3.com/image/296060239
Fold3 by Ancestry/War Diary, 5/1 – 28/44, fold3.com/image/274220824
Fold3 by Ancestry/War Diary, 5/1 – 31/45, fold3.com/image/296164137
Fold3 by Ancestry/War Diary, 6/1 – 30/44, fold3.com/image/279767147

Fold3 by Ancestry/War Diary, 6/1 – 30/45, fold3.com/image/296561292
Fold3 by Ancestry/War Diary, 7/1/44 to 8/7/44, fold3.com/image/280045389
Fold3 by Ancestry/War Diary, 7/1/45 to 8/31/45, fold3.com/image/300779272
Fold3 by Ancestry/War Diary, 8/8 – 31/44, fold3.com/image/287106685
Fold3 by Ancestry/War Diary, 9/1 – 30/44, fold3.com/image/287166622
Fold3 by Ancestry/War Diary, 9/1 – 30/45, fold3.com/image/302712891
Fold3 by Ancestry/War Diary, 10/1 – 31/44, fold3.com/image/292511763
Fold3 by Ancestry/War Diary, 10/1/45 TO 11/30/45, fold3.com/image/301881239
Fold3 by Ancestry/War Diary, 11/1 – 30/44, fold3.com/image/292544266
Fold3 by Ancestry/War Diary, 12/1 – 31/44, fold3.com/image/295215640
Fold3 by Ancestry/War History, fold3.com/image/302074195 Friedman N., *U.S. Aircraft Carriers.* (Annapolis: United States Naval Institute. 1983)
Harder, B. L., *World War II US Fast Carrier Task Force Tactics 1943-45.* (Oxford: Osprey Publishing. 2020)
Hornfischer, J. D., *The Fleet At Flood Tide* (New York: Ballantine Books. 2017)
Jacobs, R., Find a Grave Memorial/RADM Leslie Albert Kniskern, findagrave.com/memorial/60661370/leslie-albert-kniskern
Keegan, J., ed., *The Atlas of the Second World War.* (Ann Arbor: Borders Press 2003)
Lewis, C. A., Find a Grave/Adm Austin Kelvin "Artie" Doyle, findagrave.com/memorial/2935617/austin-kelvin-doyle
Marolda, E. J., *Ready Seapower A History of the U.S. Seventh Fleet.* (Washington D.C.: Navy History and Heritage Command. 2012)
Marshall, M. L. III, *Clashes. Air Combat over North Vietnam 1965-1972.* (Annapolis: Naval Institute Press. 1997)
Meredith, L. W. *Grey Ghost The Story of the Aircraft Carrier Hornet.* (Sunnyvale: Historical Indexes Publishing Co. 2001)
Miles, D., Caltech.edu/USS *Hornet* CV/CVA/CVS-12 The Grey Ghost, its.caltech.edu/~drmiles/hornet.html
Naval History and Heritage Command/Ernest E Christensen/ history.navy.mil/research/library/research-guides/modern-biographical-files-ndl/modern-bios-c/christensen-ernest-e.html
Navy History and Heritage Command/Modern Biographical Files in the Navy Department Library: Coe, Charles Frederick, history.navy.mil/research/library/research-guides/modern-biographical-files-ndl/modern-bios-c/coe-charles-frederick.html
Navy History and Heritage Command/Van Vernon Eason Jr, history.navy.mil/research/library/research-guides/modern-biographical-files-ndl/modern-bios-e/eason-van-vernon.html
Newswanger, P., Inside Business/125 Years Newport News Shipbuilding, pilotonline.com/inside-business/special-reports/article_3c5f277d-d888-53b1-906b-84bf12c84cc9.htmlPena, F., NavSource.org/NavSource Online: Aircraft Carrier Photo Archive USS *Hornet* CV-12, navsource.org/archives/02/12co.htm
Pensacola News Journal/Jackson Allison Stockton Obituary,

legacy.com/us/obituaries/pensacolanewsjournal/name/jackson-stockton-obituary?id=24722595
Raven, A., *Essex-Class Carriers.* (Annapolis: Naval Institute Press. 1988)
Roberts, G., *The Aircraft Carrier Story.* (London: Cassell & Co. 2001.)
Seaforces.org/Essex Class Carrier/ seaforces.org/usnships/cv/Essex-class.htm
Self, C., *USS Hornet A Pictorial History. (*Paducah: Turner Publishing Company.1997)
Stubblebine, D., World War II Database/Essex-class Aircraft Carrier, m.ww2db.com/ship_spec.php?ship_id=1831
The Hall of Valor Project/Ellis Jay Fisher/valor.militarytimes.com/hero/19971
The San Diego Union-Tribune/David C Richardson (1914-2015), legacy.com/us/obituaries/sandiegouniontribune/name/david-richardson-obituary?id=16775414
The W Foundation/The W Foundation, www.thewfoundation.org/
The Washington Post/Norman Gillette, Jr., retired Rear-Admiral, Aviator, Dies, washingtonpost.com/archive/local/1986/04/20/norman-gillette-jr-retired-rear-admiral-aviator-dies/2eb59476-e733-428e-aa0c-a59b8639f16f/ Toll, I. W., *The Conquering Tide War in the Pacific Islands 1942-1944.* (New York: W.W. Norton & Company. 2015)
Toll, I. W., *Twilight of the Gods War in the Western Pacific, 1944-1945.* (New York: W.W. Norton & Company. 2020)
Wikipedia/Alan Bean, en.wikipedia.org/wiki/Alan_bean
Wikipedia/Austin K Doyle, en.wikipedia.org/wiki/Austin_K._Doyle
Wikipedia/Carl J. Seiberlich,/en.wikipedia.org/wiki/Carl_J._Seiberlich
Wikipedia/Charles "Pete" Conrad,/en.wikipedia.org/wiki/Pete_Conrad
Wikipedia/Charles R Brown,/en.wikipedia.org/wiki/Charles_R._Brown
Wikipedia/HMAS Perth DD 38, en.wikipedia.org/wiki/HMAS_Perth_(D_38)
Wikipedia/Miles Browning, en.wikipedia.org/wiki/Miles_Browning
Wikipedia/Richard Gordon, en.wikipedia.org/wiki/Richard_F._Gordon_Jr.
Wikipedia/Scott Carpenter, en.wikipedia.org/wiki/Scott_Carpenter
Wikipedia/Thomas F Connolly, en.wikipedia.org/wiki/Thomas_F._Connolly
YouTube *Hornet* Museum/ Dave Pendleton: Mission Strike S-2 Tracker,
youtube.com/watch?v=-DY5_tNWX94
YouTube *Hornet* Museum/Fresnel Lens Optical Landing System (FLOS) – USS *Hornet*, youtu.be/D24JBJzLGus
YouTube *Hornet* Museum/Gene Millen WW2 USS *Hornet* Crewman Interview, youtu.be/UguNj-SwNXA
YouTube *Hornet* Museum/Tour of the USS *Hornet* Amateur Radio Club, youtu.be/mYjjLIpfeO0